MAKING A LIVING IN EUROPE

The issue of jobs is among the most important facing the European Union, where unemployment continues at near-record levels. This book focuses on changes in the economy and society that most affect employment, which is not just a specialist question: job change reshapes social inequality, lifestyles, where people live, urban congestion and environmental conflict. At the centre of the book is the shift towards flexible, part-time and temporary work for women in services. Comparisons are made between the UK, other countries of the European Union and the USA.

The central core (Part II) of the book integrates the central issue of the replacement of industrial jobs for men by more flexible jobs in the service sector for women, demonstrated by chapters on the three most relevant sectors: business services, retailing and tourism. Part I sets the European employment problem in its full global and regional context, while Part III reviews the outcome of change in terms of the shift of people and jobs from urban to rural areas and the need for stronger EU policies, not only in traditional problem regions but also to reduce inequality at large in the face of EMU.

The book explores in a challenging way some of the longstanding assumptions held about the relationships between employment change, unemployment and economic restructuring and about the effects of the shift to services. Emphasising the growth of women's employment as one result of this shift, the author questions the fashionable emphasis on 'flexible production' and uses up-to-date statistics and case material to draw out crucial points in local and national employment policies alike.

Alan R. Townsend is a Reader in Geography at the University of Durham.

MAKING A LIVING IN EUROPE

IN EUROPE

Human geographies of economic change

Alan R. Townsend

London and New York

First published 1997
by Routledge
2 Park Square, Milton Park, Abingdon, Oxon, OX14 4RN

Simultaneously published in the USA and Canada
by Routledge
270 Madison Ave, New York NY 10016

Transferred to Digital Printing 2007

Typeset in Garamond by
RefineCatch Limited, Bungay, Suffolk

British Library Cataloguing in Publication Data
A catalogue record for this book is available from the British Library

Library of Congress Cataloguing in Publication Data
Townsend, Alan R.
Making a living in Europe: human geographies of economic change/Alan R.
Townsend.
p.   cm.
Includes bibliographical references and index.
1. Labor supply – European Union countries.
2. Unemployment – European Union countries.
3. Rural–urban migration – European Union countries.
4. Europe – Economic conditions – 1945 – Regional disparities.
I. Title.
HD5764.A6T68   1997
331.1'094 – dc20   96-38677

ISBN 0-415-14479-5
0-415-14480-9 (pbk)

Publisher's Note
The publisher has gone to great lengths to ensure the quality of this reprint but
points out that some imperfections in the original may be apparent

Printed and bound by CPI Antony Rowe, Eastbourne

CONTENTS

FIGURES

FIGURES

TABLES

PREFACE

'Making a living in Europe': what will this mean in the twenty-first century? Can we learn to produce and provide for all, by one means or another? Or will 'making a living' be only a dream for many who will be deprived and excluded? As an experienced *industrial* geographer, I argue that we can only contribute by a new understanding of *services* – the book is built on research and writing on the new geographies of women's jobs and several key services over the last five years.

I ask these questions wearing different hats. I write as an *academic* with a life-long interest in who makes a living, how and where. I am fascinated by the living geography of work, particularly because of its central role in the human geography, social inclusion and exclusion, and population growth of areas. My own working life began as a regional civil servant in the struggle to attract industrial jobs to areas of high unemployment. The world has changed since then, but many economists and geographers still concentrate on activity and 'flexibilisation' in industry when most jobs in Europe now are in 'services'. In this book I shall particularly explore those sectors in which people at work experience insecurity, or 'flexibility', and ask what this means for making a living in Europe. What are the key sectors of change as seen through a broad approach to work in the European Union?

But I have another hat. Although I make my living as an academic, I am engaged in *local politics*. I live in an ex-mining community and have been elected as District Councillor and Chair of the Planning Committee for a District that lived and died from mining coal in the east and farming the Pennines in the west. Now, we struggle with a desperate need for jobs. For many, this means a crippling loss of identity. Fifteen years ago, the British government gave my local Council half the money to build in my ex-mining village the first multi-coloured artificial ski-slope in Europe, in an early recognition of the hope for jobs in the leisure industry. The Council attempted to raise £25 million for a leisure complex based on long ski slopes with artificial snow in the Pennines, but failed.

The Common Agricultural Policy, European Union resources for the rehabilitation of ex-coalfield communities and the new 'precarious' jobs are part of our everyday lives. Global change has closed the deep pits and brought us conflict over open-cast mining. Global change has shaped the jobs to which most workers in the village now commute by car. Now the Council members, who promote our District in The Netherlands as a holiday destination, and many in the village who hunger for 'men's jobs' find this both painful and ridiculous.

Everyday life in my village, as in most of Europe, must be understood in the context of the new Europe, of change in the USA, of global change, all of which I shall discuss. A particular feature of the book is that several of the early chapters consult North American experience as a warning or a guide to a continent that follows it in the search for wealth, and this is particularly relevant to one of the longest chapters, on the whole world of women's employment. I want in this book to convey to other planners and local politicians, as well as students and academics, some of what we all need to know in order to contribute positively to the struggles of local people to make a living in Europe.

Alan R. Townsend

ACKNOWLEDGEMENTS

Very many thanks are due to Dr Janet Townsend, even though we often chose to disagree on academic and policy matters, my father Cyril Townsend, and to Joan Andrews, Lisa Tempest, David Hume, Anne Stokes, Dr Tony Champion, Dr Neil Coe, Dr Nick Cox, Prof. Peter Daniels, Jim Lewis, Dr Rachel Macdonald, Dr Neill Marshall and Dr Joe Painter for help and encouragement of many kinds.

The author and publisher also wish to thank the following for permission to reproduce copyright material. Every effort has been made to contact copyright holders and we apologise for any inadvertent omissions.

Roger Berry MP for extract from *Economic Policies for Full Employment and the Welfare State*; Blackwell Publishers for extracts from R.J. Johnston and P.J. Taylor and R.E. Pahl; Cambridge University Press for extract from H. Giersch, K. Paque and H. Schmieding; Carfax Publishing Company, PO Box 25, Abingdon, Oxfordshire OX14 3UE for extracts from articles by J. Bachtler and R. Michie in *Regional Studies* 27, 28, 29; CEDEFOP for extract from *Occupations in the Tourist Sector* (1994); Commission of the European Communities for extracts from various publications; *The Economist*, London for three extracts and Figures 2.6, 2.8, 5.1 and 5.2; Equal Opportunities Commission for extract from *Labour Market Structures and Prospects for Women* (1994); *The Guardian* for extracts from articles by C. Brewster and B. Laurence; *Financial Times* for five extracts and Figures 4.2 and 11.2; Heinemann Educational, a division of Reed Educational and Professional Publishing Ltd, for extracts from D. Bell, *The Coming of Post-Industrial Society* and F. Blackaby, *Deindustrialisation*; International Labour Organization for figure from *World Employment 1995. An ILO Report* (1995), p. 30; Kreisky Commission for extract from *A Programme for Full Employment in the 1990s*; Longman for extract from J.N. Marshall and P.A. Wood; M.E. Sharpe, Inc., Armonk, NY 10504 for extract from A. Etzioni (*et al.*), *Socio-Economics: Towards a New Synthesis*; Macmillan Press Ltd for extract from D. Gregory, R. Martin and G. Smith; MIT Press for extract from J.H. Dreze and C.R. Bean; The Open University for Figure 2.1, adapted from A. Thomas *et al.*, *Third World Atlas*

(1994), second edition; Oxford University Press for extract from the *Oxford Review of Economic Policy* 11; Pion Ltd, London for extracts from articles appearing in *Environment and Planning A* (1990) 22, 1337–54, (1992) 24, 1255–70, (1994) 26, 1397–1418, (1996) 28, 1843–58; Routledge for extracts from S. Wood, C. Harvie, and A. Amin and J. Tomaney; Royal Geographical Society for extract from *Western Europe: Challenge and Change* (1990); Simon & Schuster for extract from R. Reich, *The Work of Nations* (1991); The Social Market Foundation for extract from M. Symonds; UCL Press for extract from R.D.F. Bromley and C.J. Thomas; West Virginia University for extracts from *International Regional Science Review* 12 and 16; Princeton University Press for extract from G. Esping-Andersen.

CONVENTIONS FOR THE EUROPEAN UNION (EU)

The contemporary usage, referring to the European Union as EU, is used throughout, except when older quotations may refer to the 'European Community' (EC).

Following the practice of many EU publications, the acronyms 'E12', 'E15', etc. are used to refer to the number of countries at different stages of assembling the Union:

E6 (1957) Belgium, France, Federal Germany, Italy, Luxembourg,
 The Netherlands
with additions to comprise:
E9 (1973) Denmark, Ireland, United Kingdom
E10 (1981) Greece
E12 (1986) Portugal, Spain
E15 (1995) Austria, Finland, Sweden

This volume includes certain data for E15, but not all statistical series have been retrospectively revised by Eurostat, in fact some countries that joined earlier have still not all supplied standardised data to Eurostat publications. Regrettably, therefore, comparison is not possible between all the tables.

Despite considerable variation in the attribution of EU publications, all are standardised in this volume under 'CEC' (Commission for the European Communities', usually published in Brussels) or 'CEC, Eurostat' (usually published in Luxembourg).

To provide for standardised comparison with the rest of the industrialised world, data are used from OECD, the Organisation for Economic Co-operation and Development, Paris, which in tables available at the time of writing comprised the above countries (E15) plus Canada, the USA, Japan, Australia, New Zealand, Iceland, Norway, Switzerland and Turkey.

Part I

RESTRUCTURING, FLEXIBILITY AND UNEMPLOYMENT

1

PRINCIPAL THEMES

Making a living is at the centre of our lives. Jobs are one of Europe's most important topics, with their obverse of unemployment, or increasingly 'non-employment', when people of working age have simply stopped searching for work. Employment provides the basic means of distributing wealth in society, in providing income for young families and ensuring pensions for elderly people. Paid work is the basis of leisure spending or a town's provision of services of all kinds.

This book explores and challenges some longstanding assumptions made by geographers and others about jobs. At the core of the work are the changes in the economy that most affect employment. Gone are the days when economic geographers could comfortably concentrate on the unionised work of men in manual factory work – that is no longer typical. The book integrates in one text the central issues of change: the loss of factory jobs ('deindustrialisation'; Chapter 4), the relative growth of service jobs ('tertiarisation'; Chapter 5) and the growth of women's jobs ('feminisation'; Chapter 6). Further chapters explore in more depth the three most relevant sectors, the two that consistently show the strongest growth of jobs ('producer services'; Chapter 7 and tourism; Chapter 9) and one chosen for massive change in the quality of jobs (retailing; Chapter 8).

These changes, providing the core, Part II, of the volume, shape the overall total of jobs and so both produce and counter the growth of unemployment and low pay in individual types of area. As the European Union is increasingly a single functioning unit for economic purposes, with many increasing similarities between member states, it is chosen to provide the framework in which the study proceeds. The sectoral studies are therefore prefaced by a base-line study of the problems of employment across *all* sectors in the EU area, Part I, while the outcomes of the sectoral studies across different types of area (urban to rural) are assessed, with overall implications for the EU as a whole, in Part III.

This first chapter will set the context in human geography for the principal themes of the book, 'sectoral change' and 'flexibilisation'. Chapter 2 will introduce the implications of unemployment, and Chapter 3 will undertake a spatial approach to Europe's position in global economic trends.

The redistribution of work has changed relations between men and women. The culture of work in industrialised countries has been transformed through reduction in the regular demand for men's manual work and change in women's time-budgets to include paid work. This is amply recorded in the discourse of the first of several regions that we visit.

A voice from the past

The day after I left school I went straight down the pit, with most of my mates. I had eleven good year down Brancepeth Colliery before it closed in 1964. The Coal Board gave me a job six mile away at Brandon, before it too closed – hitting a fault they said but we all knew this area had had it. I was lucky that I knew Jimmy Knight running wagons for Hunwick Brickworks, and I did all right there for a couple of year before he went bust and I went on the dole. My sister-in-law made enquiries and I got a job with the Council bin-men. That was tough work, but reliable, until these Tories made them put the work out to tender. The union thought it might be all right, but I was one of them made redundant, with only a little compensation. That were five year ago; I'd had some back strain, and I went on the invalidity. Best thing that could have happened really, when I was 53, because there's no future round here for anyone over 40 – if that, and you see all these youngsters roaming round at night with nothing to go to in the morning; should get out like my son who's a teacher down London.

(Member of Willington Working Men's Club,
Co. Durham, England, 1995)

The speaker is typically silent about his wife, who will probably survive him as a widow. Willington is my home, in one of the longest recognised UK problem areas. I am a District Councillor and Chair of the Planning Committee as well as a geographer, which defines my interests in change both in Willington and Europe. Our old primary industry (in this case mining) has been succeeded not by secondary (manufacturing) industries but by *an uncertain mixture of tertiary (service) jobs and underemployment* – a familiar tale across Europe. The men's jobs at the big colliery went when it closed in 1964 and our 5,400 villagers now have only 820 local jobs, occupied mainly by women. Those men in jobs travel to work across much of County Durham. However, paid work for men is increasingly confined to the prime age groups, as nearly a quarter of men are still unemployed between 21 and 24, and half have left the labour market at 55–64. A quarter of the whole population reported a 'limiting long-term illness' to the 1991 Census of Population.

At the end of this chapter comes a comparable tale from Southern Italy, another classic problem region, although the details are completely different.

Longstanding migration from the land was only partly met by past industrial-isation policies, and growth of service jobs has failed to prevent continuing high unemployment, in this case especially among young women. In the 1990s high unemployment has spread to many new areas of the EU, notably in Germany and Scandinavia.

My journeys between North-East England and other areas of the EU such as Denmark or Southern Italy take me into totally different realms of cultural geography. These regions have reached their present human geog-raphies by different routes, and the influence of different inherited social, cultural and political practices on the pattern of jobs is strong. However, two points arise.

First, the reality of the Working Men's Club is partly captured by the local job statistics. In understanding human experience, statistics of employment can help us compare regions and countries of the European Union (EU), using due care and attention to the dangers of the data.

Second, whatever the cultural and historical contrasts within the EU and whatever the dangers of sweeping social science 'laws', different regions are, in their experience of making a living, facing similar constellations of trends – such as redundancy or service employment growth – in their *directions of employment change*.

MAIN THEMES

My aim in this book is to understand the future of *places* in industrialised countries, in terms of these directions of employment and economic change at large. These prospects depend on the EU and its members' national economies as a whole. More tangibly, however, they turn on the *particular mix of sectors* found in the individual place, perhaps in business services, tourism or manufacturing. The *changing division of labour* by sector, gender and place is my main theme, with the *more flexible uses of labour* in a society more oriented to consumption, leisure and services. The context is of unemployment and non-employment as they exist and as they enter future policies of the EU.

The future of individual places also hinges on regional planning by the EU, national and local governments, and more fashionably on community initiatives in local development and employment (CEC, 1995a). All member states evolved measures by which they attempt to correct unemployment and weakness in the economies of constituent regions and sub-regions. Not only do these schemes make additional finance available to attract companies but they may invest in transport, water supplies, training facilities or telecom-munications. All areas will normally make planning provision and forecasts for their future needs, say in terms of the roads or industrial land to be created in a given forward period. A 'jobs gap' used to be calculated for this purpose, by assuming that decline in particular sectors would have an effect on unemployment and that nearly all men would continue to seek work. Such assessments must now include the quality of work, threatened as this is by strong tendencies toward 'flexibilisation'.

Two kinds of restructuring may be seen in Europe.

Sectoral change occurs as service jobs replace factory, mining and farm work. Few areas still have an expanding total of manufacturing jobs. Industrialisation was seen as the 'solution' to local and regional problems. Now we experience 'deindustrialisation', the net loss of industrial jobs or of export production, and too little thinking about possible futures. I agree with Martin (1994: 3): remarkably, even the most fashionable theories have 'barely recognised the service-based regional economies' of most European capital regions. I shall not only explore the 'withering' of the old conventional jobs, but ask where continuing jobs will be. I shall compare the human effects of change in important sub-sectors of the economy such as financial and producer services, retailing and tourism. Whatever the starting points in different areas, the universal *direction of change* in the industrialised world is toward services, women's and less conventional jobs, and often from urban to rural areas, 'counter-urbanisation' (Champion, 1989).

Flexibilisation comes about in many ways in different sectors as employers seek to remain competitive or efficient through adapting the volume and content of their work to changing demand. Workers in part-time, temporary and self-employed work may take over from those in safe, full-time jobs, which accounts for much of the new levels of fear felt widely by employees in the 1990s. The EU is not uniform. 'Formal' work has never been as important in Southern as in northern Europe (Stratigaki and Vaiou, 1994). However, the whole future of the EU is indissolubly bound up with the quality of jobs that are being generated and with the continuing problem of unemployment.

PAST ECONOMIC GEOGRAPHY – WAS IT HUMAN?

I have chosen deliberately the sub-title 'human geographies of economic change'. To pursue the above task we must start by appreciating, even in the 1990s, that economic geographers have tended both to ignore the human aspects of change and to explore theories only for the manufacturing sector.

The principal reason for studying the location and growth of economic activities must be for their effects on daily life, all the more critical in a world of underemployment. In economic geography, human beings have often been silent. Much commercial geography dealt exclusively with world commodity distributions and international trade. Labour market studies were normally only a sub-set of industrial geography. Labour supply achieved merely a recognised deviation from the optimum location of an industry as modelled by Weber (1929), and played no part as such in the work of Loesch (1954). It was only during the 'long boom' of expansion after the Second World War, when labour became extremely scarce in an era of full employment in much of western Europe, that labour attained a premium as an *input* to industrial location decisions. Such decisions could, however, be influenced through government regional policies, as in the UK or Southern Italy, to redress regional imbalances, often crudely measured through unemployment rates for the filling of the 'jobs gap' by new vacancies of whatever kind.

Massey (1984) deployed Marxist concepts to display a 'spatial division of labour' in the exploitation of space by multi-site industrial corporations. A single corporation might rely on established skilled labour at one plant in a northern French industrial town, on routine work from a former agricultural workforce at a plant established for the purpose in Brittany, on available graduate staff in an environmentally attractive area such as the French Alps for research and development and laboratories, and on well-trained administrative staff in Paris for head office functions. Each of these functions would exploit the past role of the area and in turn shape its class and income distribution. Massey showed how certain areas were incorporated only for their semi-skilled labour, including many areas to which industry spread in Southern Europe in the 1960s and 1970s. Many of these same corporations then made use of labour in low income countries, contributing to a 'Global Shift' of industrial output (Dicken, 1986).

The selective 'downsizing' of these corporations in more prosperous countries through massive job cuts (Townsend and Peck, 1985) was accompanied by two tendencies in research. Those geographers who turned to the study of small firms focused '. . . not directly on small firms as people with psychological and social characteristics. Instead the tendency has been to treat employment as net additions or deductions from an abstract entity – the labour force' (Curran and Stanworth, 1986). Mainstream geographers stressed employers' use of plant and labour in more flexible ways, a valid approach save that they concentrated almost exclusively on the industrial economy when those jobs were declining and the concept applied to many more people in the service sector.

Labour market questions may be best seen only as a subordinate part, a local end-product, of change. Much analysis of the location of services has always and rightly been focused on market issues rather than any considerations of labour supply. Domination of the labour market by services is nevertheless of great consequence, as seen in the US writings of Hanson and Pratt (1995) on gender, work and space in contemporary Worcester, Massachusetts. Most industrial geographers diversified their work into the service sector after industry contracted (e.g. Daniels, 1985; Law, 1994; Shaw and Williams, 1988). Yet more Geography Option Course topics appeared, in Business and Financial Services and in Tourism, without anyone releasing their joint potential in studying the overall modern system of economy, politics and daily movement.

Labour remains an important *input* to location decisions. Cheap, docile labour remains more important than the existence of weak environmental law in attracting fresh waves of global industrial investment into lower income countries. Skilled labour remains the key feature in the production and reproduction of the world's financial centres, and tourism, depending on a wide range of labour, can be seen to be restrained in areas of congestion and scarcer labour.

Today, we need a people-based economic component of human geography, rather than 'economic geography'. It would draw on early work such as Pahl (1984) on *Divisions of Labour*, and Redclift and Mingione (1985), *Beyond Employment*, extended to Sayer and Walker (1992) on the *New Social Economy*. It would reflect the replacement of men's work in large corporations by a more complex mix of work and leisure activity, involving many forms of less conventional work in a mass of more specialised small firms. This would also reflect the ascendancy of a broader human geography over the narrowly economic in the 1990s, and a fresh move to reconcile social and economic approaches in social science as a whole.

Socio-economics

Dissatisfaction with the apparently inhuman nature of standard economics has brought new writing about the structure of economic decisions and their effects, which has important parallel implications for geography and some for employment:

> Socio-economics was born of a need to understand man [sic] and how hard he works, how much he saves and what he purchases ... socio-economics assumes the decision-making units are individuals integrated into one or more social groups, ethnic groups and subcultures.
>
> (Etzioni, 1991: 3)

Likewise Sayer and Walker (1992) hail the 1990s rediscovery of the social nature of economies, involving the multiplication of many kinds of activity especially in the service sector.

In place of economic forces, socio-economics places power relations between individuals and groups at the centre of economic change. Analysis thus centres on entrepreneurial activity and competition, seen as social activities of individuals in firms, partnerships, co-operatives and large corporations, but the concern with institutions is extended to embrace regulations, rules, habits and conventions such as language (Matzner and Streeck, 1991), taking us into cultural geography.

Sayer and others have set limits to the 'cultural turn'. There are still two leading ways in which socio-economics can interact with modern human geography. We must leave behind the 'under-socialisation' of economic models (Granovetter, 1985) and examine how successful firms relate to their surrounding social and political milieu, how they are 'embedded' in their context, their geographical milieu. Corporations like Volvo are embedded in a culture, and the surviving plants of large corporations may be more 'embedded' in their areas than before; new firms in Central Italy have benefited from a set of social, cultural and political relations providing a successful environment for growth.

The cultural context of employment

Beyond this, the amount of employment itself may well be a cultural phenomenon, in terms both of different practices inherited from Europe's nation states and of the changing demand for work today. The quantity of work demanded in a society may no longer be a fixed or economically determined quantity, as under post-war conditions and assumptions; it may be culturally determined, for instance by women's changing practices in the household:

> The recent breakdown of most cultural and legal barriers to access to the labour market in most Western countries makes it immeasurably more difficult to define the target of full employment policy. Where high female participation rates, for example, are accorded greater social value, social and cultural change vastly increases the burden on full employment policy. . . . As far as comparative analyses of employment performance are concerned, taking into account the obvious differences in cultural values that are at least partly responsible for the differences in female participation rates between, say, Sweden and The Netherlands becomes a difficult and challenging problem.
>
> (Matzner and Streeck, 1991: 5)

The diversity of gender relations and thus of women's employment across the cultures of western Europe (Duncan, 1994) is more than a statistical quirk. (In this sense our work is linked with the renewed recognition of national identity and its reflection in political and cultural geography.) The reconstruction of labour supply has made it possible to expand employment despite low growth in productivity and even at declining real wages, with more members of the household seeking work as wages go down. This experience of the USA in the 1980s reminds us that a widespread growth of service jobs may be much more important to human lives than investment in capital-intensive manufacturing. Significant as it is to the national economy, an efficient petrochemicals plant with few workers is not very important to its local economy if it does not make many purchases in the area.

In Western societies, it is possible to produce enough for all, but not be able to distribute the wealth adequately because wage work remains our leading mechanism for distributing the proceeds of production to able adults. A premise of mine is that there is little point in solving a nation's economic problems if reasonably full employment is not also restored in all geographical areas of the country, or indeed if the environment is not protected from damage.

SECTORAL CHANGE

In most accounts, sectoral change and, more recently, flexibilisation do dominate, and I shall now take each in turn. In my view the academic world has been obtuse in its failure to accord more recognition to services. Europe illustrates the immense scale and spatial concentration of growth in service employment. The human and economic significance of office expansion is very evident in the Randstad (Amsterdam, The Hague, Utrecht and Rotterdam), Stockholm, Copenhagen, Frankfurt, Paris or London. The quality of geographical work on such development has matured through setting social linkages, as above, in a context of changing metropolitan demand.

But can studies of all the different parts of the service sector, from international finance to hairdressing, be integrated? Does it have any overall characteristics or dynamic? How does it relate back to sectors of production? The distinction between services and other sectors is very important for conceptualising how individual places might 'make a living in Europe'. But I have gained little from trying to generalise the characteristics of different parts of the service sector, which all have different relations with the rest of the economy. I would instead follow Marshall and Wood (1995) in proposing a 'service-informed' view of geographical change, rather than a single, monolithic view of the sector.

International service expansion

The need to recognise services is made even more critical by the faster pace of structural change in the last twenty years. The balance of jobs in E12 has critically changed, with services claiming 63.5 per cent of employment in 1992 compared with 46.6 in 1972 (OECD, 1994a) (or 61.8 in E15, 1992 and 63.9 in 1994; CEC, 1995b). The pace and direction of change are similar in very different regions of the EU (Figure 1.1). From the service-sector share of total employment attained by the end-dates used in the figure, it is clear that the levels attained in Northern England, far from being peculiar to an exceptionally deindustrialised region, were being approached by, for instance, West Berlin and Campania, following marked shifts in the 1980s. Industrial areas such as the Flanders area of Belgium (which excludes metropolitan Brussels) also showed a decisive shift from industrial employment. The areas of greatest industrial strength, many located 'around' the Alps, did seem to prosper without needing a strong growth of service activities, but these regions, including Bavaria and Rhone-Alpes (Chapter 3) have since shown a decisive shift towards a preponderance of service jobs.

Technically, Marshall and Wood (1995: 21) are right in saying that 'there is

10

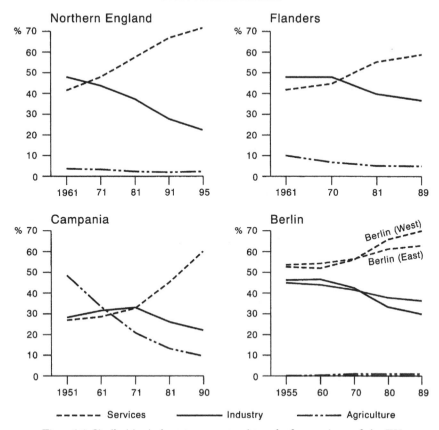

Figure 1.1 Similarities in long-term sectoral trends, four regions of the EU:
percentage shares of total employment
Source: CEC, 1993a

no uniform shift towards a service-led economy'. Different countries, with their different cultures and politics, have had different mixes of services involved in employment expansion in different historically specific decades. Nonetheless, all parts of the sector have expanded in the EU, under whichever Eurostat head of distributive trades, hotels, transport and communication, banking, public administration or finance and insurance.

Figure 1.2 displays the EU countries as part of an inter-continental pattern of employment transfer between agriculture, industry and services. Each country has two intersects on the graph, for dates in the 1960s and the 1980s respectively, and change between the two dates is shown by joining them by an arrow. In all parts of the globe, not least in agriculturally dominated economies, the proportion of workers in primary industries (not necessarily the total number) is giving way to other sectors; this shift out of agriculture has been important for people's lives and for regional

11

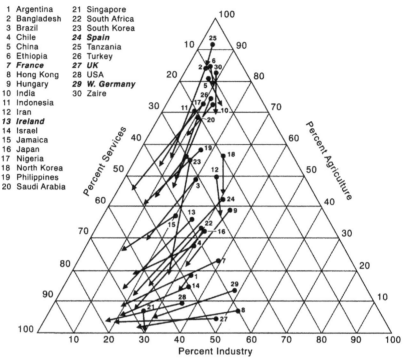

1 Argentina	21 Singapore
2 Bangladesh	22 South Africa
3 Brazil	23 South Korea
4 Chile	24 *Spain*
5 China	25 Tanzania
6 Ethiopia	26 Turkey
7 *France*	27 *UK*
8 Hong Kong	28 USA
9 Hungary	29 *W. Germany*
10 India	30 Zaire
11 Indonesia	
12 Iran	
13 *Ireland*	
14 Israel	
15 Jamaica	
16 Japan	
17 Nigeria	
18 North Korea	
19 Philippines	
20 Saudi Arabia	

Figure 1.2 World convergence towards the dominance of service employment: change in sectoral distribution of labour force for selected countries, 1960/5 (circle) to 1986/9 (arrow head)
Note: EU countries are in bold italics
Source: Thomas *et al.*, 1994

development in parts of Europe within the period of the graph. In turn, however, industrial jobs are giving way to service work as most countries take a decisive turn for the bottom left-hand corner of the graph. There is convergence in the sectoral direction of change.

'Explanation' of the shift toward services

The conventional Fisher/Clark hypothesis envisages progressive shifts from primary to secondary and then to tertiary (service) activity at successively higher income levels per head (Fisher, 1935; Clark, 1940). Descriptively there is not much scope for argument about this! But what causes the relative growth of services?

With higher incomes the composition of household demand certainly takes on a different structure, with proportionately less spent on basic necessities of housing, food and heating and relatively more on leisure, financial and government services. The over-riding factor, however, has been viewed as growth in productivity. On this hypothesis, a high income society achieves high levels of production per worker through automation in manufacturing, whereas in many service activities such productivity is difficult to increase; good examples are nursing, teaching or hairdressing. As the economy grows, any increment in manufacturing activity may be accompanied by job loss, whereas the resulting income is spent on services that can expand only by increasing their workforce. The balance of employment thus shifts towards low productivity services.

There are a number of criticisms of this position. Gershuny (1979) argued that rising income had principally stimulated the demand for goods, rather than services, and that people were substituting manufactured household appliances and DIY equipment for service provision. Others argue that productivity is difficult to measure in services but that there are parts of the service sector that have held down employment levels despite large increases of throughput. In fact 'Any blanket statement about the relative productivity of services and its employment implications is obviously unrealistic' (Marshall and Wood, 1995: 17).

The importance of services in their own right

Still, 'many economists and geographers continue to deny that the enormous shift towards services has changed any of the forces and categories fundamental to industrialism' (Martin, 1994: 34) because they see the essential basis of wealth creation as lying in the material processing, extractive and manufacturing sectors. More of the work of production has indeed been transferred to fall technically under the heading of service firms and industries, to design and computer consultants, to advertising and legal agents, or to catering, security or transport suppliers (Sayer and Walker, 1992). There is ample evidence that 'lean production' in the 1980s has been accompanied by the increased contracting out of services, and that this yielded a significant increase in 'service' jobs.

However, this is another conservative way of thinking. We now know (i) that only a minority of 'business services' are actually for the manufacturing sector (only 10 per cent in a survey of Edinburgh undertaken by Townsend and Macdonald, 1995); and (ii) that these business services represent only about a quarter of the service sector.

The traditional view expressed by Sayer and Walker (1992), that most of the service sector is merely an adjunct to production in other sectors, can be stood on its head. Many jobs in 'production' are in service occupations. The non-production activities of manufacturing corporations, in transport, distribution, research and development, administration and head offices are detached from the factory locations and occupy freestanding premises. *This is of the utmost importance for regional development and local planning bodies,* as the firms may choose their locations for development independently of the manufacturing operation.

Even more importantly, considerable parts of the service sector proper, including larger public and private offices, are located independently. Thus, to understand the availability of work for a specific geographical area, we need to refer not just to the dynamics of industrial location but to the opportunities for investment by service activities. In the service sector proper, a large proportion of activity is associated with the distribution of population – but much is not.

> It is not just that services now outweigh manufacturing in the economy; it is also that in some cases the distinction between the two is increasingly blurred, while in others services have become an autonomous source of growth, demand, capital accumulation and economic regulation, no longer simply linked to or dependent on industrial growth but having their own structural dynamic.
>
> (Martin, 1994: 26)

Townsend (1991a) argued that the scope for adding activities to a local economy, because they brought in money from outside, extended well beyond the field of business services to include larger government offices, tourism of all kinds and longer distance retailing. Thus the sectoral shift towards services may bring disproportionate gains for some places, and relative losses for others. The potential gains should be considered by all regional and local authorities in Europe. This is subject to one vital consideration, that income per head, and conditions of employment, *may* be worse in the service sector than in others (Townsend, 1991a). That is because it was the pioneer in the process of the 'flexibilisation' of employment, the subject to which we now turn.

FLEXIBILISATION OF WORK

Study of 'flexibilisation' must accompany and inform sectoral analysis as a fundamental change inflicted by employers on employees, sometimes through making them 'self-employed'. We know from our own family and friends that the end of the 9 a.m.–5 p.m. job for life, long heralded, is appreciably closer. 'Flexibilisation' of employment creates the opposite of the old work pattern. Employers may vary numbers of staff by the hour, day or week or use temporary contracts over longer periods. Traditional full-time jobs are fewer and feelings of job insecurity are widespread in different national conditions. The 'flexibilisation' of the economy is thus of immediate importance.

The starting point is commonly taken to lie in the mass production techniques of transnational and other industrial corporations, normally described as 'Fordism' to reflect the early techniques used by the Ford Motor

Car Company in streamlining its production lines. The pressure of declining manufacturing profits in the 1980s led to painful restructuring through closures and redundancies (Townsend, 1983). Increased international competition brought greater pressure to meet more sophisticated consumer choice through varying main product lines over short periods of time. Coupled with still greater pressure for productivity improvements, a regime of 'flexible specialisation' (Piore and Sabel, 1984) was described as generating networks of independent suppliers in individual growth areas. Some of these suggested areas are located in the EU, including the Third Italy between the north and south.

Academics thought that changes in employment and labour practices were necessary for advanced capitalist economies to overcome the economic crisis that characterised the early 1980s. Attention was particularly given to two types of flexibility. Firstly, what Atkinson and Meager (Atkinson, 1985, 1988; NEDO, 1986) call *numerical flexibility* or Streeck (1987) calls external flexibility: the enhancing of the firm's ability to adjust labour inputs to fluctuations in demand. Second, *functional* (or internal) *flexibility* refers to workers' actual tasks and workers' ability and readiness to be transferred between tasks according to the firm's workload, production methods and technology. Allied to both is *pay flexibility*, under which a firm can adjust labour costs, particularly pay, to changing market conditions.

The application of 'flexibilisation' to industry or services?

Two questions are important to me: firstly, whether 'flexibilisation' is more important in industry or services, and secondly, whether it is indeed advancing. On the first question I feel that an answer can be given now, with the support of other researchers. The great majority of industrial geographers were stuck in a timewarp:

> The emphasis on physical production is ... bizarre in the sense that many of the characteristics of the new more-flexible operations described have for a long time been prominent in services, for example, in the importance of flexible management styles, part time employment, small firms, vertical disintegration, intense local interaction, and heavy investment in information technology.
>
> (Marshall and Wood, 1992: 1262)

I also agree with Martin (1994: 31–3) that 'the claim that flexible accumulation is replacing mass production is questionable' and that 'not enough concrete examples exist to support the general proposition advanced by the theory'. In a general sense I conclude that the question of flexibility in labour deployment has been much discussed in the industrial sector but mainly practised in services.

But is it increasing? Here I can come to only a preliminary view at this

15

stage. In considering *The Transformation of Work,* Wood (1989) argues that the precise form of the flexibility debate has varied among countries, depending on the prevailing types of flexibility:

> Some element of flexibility in work organizations has always existed, and it is doubtful if a rigid assembly-style mass-production system was ever dominant enough to warrant its being the central organizing principle of models of industrialism. The current impact of technology and new managerial concepts will vary depending on the starting point, the sector and even the firm. The overall message of the book though is that there is no reason to suppose that changes in work organization are all headed in the same direction.
>
> (Wood, 1989: 43)

Initial evidence

The USA and UK governments brandish movement to a flexible labour market as a virtue for competitiveness. By 1996 there is more evidence of change and the European Labour Force Survey is now more equipped to report it. The UK has a more diverse pattern of hours than any other EU member state (Beatson, 1995). Nonetheless, other European countries were catching up from 1983 on the UK as regards their proportion of workers who were part-time (Naylor, 1994). The overall EU average change from 1988 to 1994 is shown in Figure 1.3, demonstrating a marked shift in the growth of jobs from full time to part time in the recession conditions of the

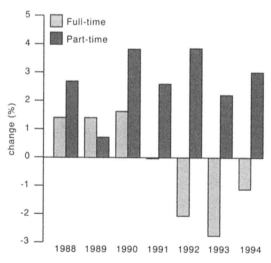

Figure 1.3 Growth of full- and part-time jobs, per cent, E15, 1988–94
Source: CEC, 1995b

16

first part of the 1990s, implying the actual flexibilisation of some jobs through full-time losses being replaced by part-time gains.

The European Labour Force Survey also enables us to sum the proportion of part-time workers, at Table 1.1, with self-employed, temporary and family workers, all the 'flexible' groups we can find in Eurostat data. Although there is a little overlap between these groups, this produces a total of no less than 38.5 per cent of European employment as falling, by 1995, outside the confines of full-time employee contracts. In addition, we glimpse weekly variations in hours worked, with nearly 18 per cent of workers significantly varying their hours of work in a given sample week, which is primarily connected with the incidence of overtime and holidays.

European work is based on a much more varied and flexible time-space weekly diary than had previously been thought. The overall trend appears to be towards greater flexibility, but at different rates on different measures and even then somewhat more gradually than would be imagined from the level of academic and political discussion. This is partly because the growth of part-time work, certainly among women in the service sector, was well established *before* this discussion took place.

A new kind of economic geography

Diversification in the form of earnings has further implications for the economic functioning of individual areas. The reduction in full-time formal

Table 1.1 Flexible forms of European work, 1992 (thousands)

Persons in employment, E12	140,175	100%
of which		
Self-employed and employers	21,249	15.2%
Family workers	3,332	2.4%
Part-time employees, of which	16,906	12.1%
women	14,408	
service sector	14,124	
not wanting full-time work	11,879	
Temporary workers	12,354	8.8%
In a given week, worked		
more than usual hours	10,203	9.8%
less than usual hours	17,677	12.6%
Usually work shifts	13,602	9.7%
Usually work evenings	14,328	10.2%
Usually work nights	7,122	5.1%
Usually work Saturdays	38,856	27.7%
Usually work Sundays	14,940	10.1%
Usually work from home	6,896	4.9%
Holding two jobs	3,719	2.7%

Source: CEC, Eurostat, 1994

work, at least in northern Europe, during the process of restructuring since the mid-1970s has meant that a greater proportion of an area's income arises from self-employment, informal earnings, and tourist spending, together with unearned investment income and welfare payments. In terms of economic geography, this has meant that the conventional model of the local economic base, which previously assumed that an area made its living in the world by exporting material goods, has changed at the same time as the national model. Factories and full-time work have become relatively less important in the economic, social, cultural and political life of the average town, at least in the north of Europe. Arguably this produces convergence with regions of Southern Europe which had been slow in moving towards a formal structure of full-time employee work. In fact, the regional case study that follows provides a sad tale of longstanding flexible employment persisting in an area of relative poverty.

SECTORAL CHANGE AND FLEXIBILISATION – AN EXAMPLE FROM SOUTHERN EUROPE

The distance between the human scale and that of formal sector external investment is illustrated by the familiar example of restructuring in the very distinct cultural area of Southern Italy (Figure 1.4). Since the 1950s, international theorists of regional development have focused on it, as since 1957 has the then European Economic Community as its worst constituent problem area, from its foundation in the Treaty of Rome (1957). Yet how many readers know that the highest unemployment of this region today is among

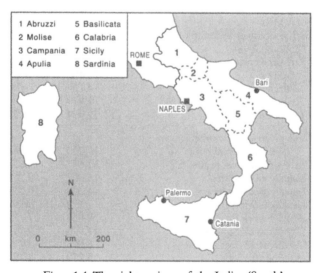

Figure 1.4 The eight regions of the Italian 'South'

18

young people, women and graduates? Italy is now on average better off than Britain, but its employment problems have stubbornly changed shape, not disappeared.

The area is very notable for 'flexible' employment. At one extreme, a large number of young people in areas such as Naples are working in the 'informal sector', that is without contracts, in activities such as door-to-door sales, giving private lessons, childminding and typing, while men may be involved in hairdressing, hawking or decorating, and women employed at home making leather goods, paper flowers or toys (Mingione and Morlicchio, 1993). Some of this work is illegal: workers pay no tax and enjoy no legal protection, security or workers' benefits. The pathways to escape are limited by an inefficient school system and inadequate vocational training, while the social welfare net is coarse, complicated and uneven. The support of the traditional family system has been weakened by the increasing separation of parents (though in later decades than in other industrialised countries), and the poor employment conditions, unemployment and weak state support combine in a process that leads to multiple deprivation and crime. Mingione (1988: 562) explains how Mafia activity can involve thousands of people, while youth unemployment provides 'a very useful "reserve army" of labour for organised crime'.

In Naples (Figure 1.4), more than a quarter of the population are officially calculated as being 'in poverty'. Nearly a quarter of people of working age are out of work; this region as a whole had a 1995 unemployment rate of 25.9 per cent (the highest in Italy), far exceeded by the average for women of 34.8 per cent and no less than 68.7 per cent for young people under the age of 25 (CEC, Eurostat, 1996a). Campania was described by European officials as having 'a situation which is ridden with problems, chronic unemployment, mounting organized crime, glaring inequalities between social classes and a persistent self-interest among the ruling classes' (CEC, 1993a: 228)

If we adduce structural employment change as part of the explanation, we find that development efforts of the 1970s achieved a level of nearly 40 per cent of workers in industry, but that by the 1990s the proportion had fallen back below the level of 27 per cent pertaining in 1951. Industrial job loss has combined with the long-established migration from the land to leave over two-thirds of workers in the burgeoning service sector, itself unable to absorb the effects of the highest birth rate in Italy.

With varying emphases this pattern of change applies to the rest of Southern Italy. Extensive regional policies since 1950 for agriculture, industry, tourism and the infrastructure appear to have failed. Taking the E15 average as 100, the value of production (in GDP per head) from the three poorest regions of Italy in 1993 stood at 69 (Campania), 66 (Basilicata) and 60 (Calabria), whereas Italy as a whole stood two points above, and its wealthiest region, Lombardy, stood at 31 per cent above the European average per capita (CEC, Eurostat, 1996b). Although these figures for the South show deterioration, especially compared with the vibrant industrialisation of the 'Third Italy' in areas around Bologna and Florence, nonetheless there has been a southward transfer of wealth through national subsidies and welfare benefits (King, 1992). Yet in one of Europe's leading countries we see the persistence of broad areas of poverty and social exclusion, increasingly in metropolises.

SUMMARY

- The employment question is central to the future of the EU. In many countries, changes are taking place in similar directions; for example, whatever their starting point, all countries are experiencing increases in the share of women in employment.
- Two kinds of restructuring are evident: a new distribution of jobs between sectors and a greater 'flexibilisation'.
- A more human approach to economic activity must include awareness of cultural variation and change. This lays bare the fluidity of whole concepts such as 'full employment'.
- Services, the sector with most job increases, are extremely diverse, but several can contribute to the local or regional economic base. Promotion of these may pay off in jobs.
- The example of Southern Italy illustrates how Southern Europe may remain significantly different from the rest of the EU in the quality of social experience and the quantity of work (one theme of the book).

ABOUT THIS BOOK

I want to provide an integrated basis for a better understanding of present-day socio-economic life in the European Union. Having established, in the context of other past and present approaches, two main themes, of sectoral change and flexibilisation, I am concerned with those branches of employment that most *produce* unemployment by losing jobs or most *offset* it by creating them.

To identify these branches, Figure 1.5 exhibits the recent percentage change of employment in different sectors across E15, and identifies domains of loss and expansion for analysis in Part II. In manufacturing, although a few industries managed a little growth from 1985 to 1990, the great majority (especially heavy industries) lost jobs at a high rate, so Chapter 4 explores 'deindustrialisation'. Chapter 5 examines services as a whole, particularly their flexibilisation, in the light of trends in the USA. Chapter numbers on the graph then identify individual services to be investigated – finance, insurance and business services (notable for their strong growth) in Chapter 7, retailing and wholesaling in Chapter 8 (partly because of their

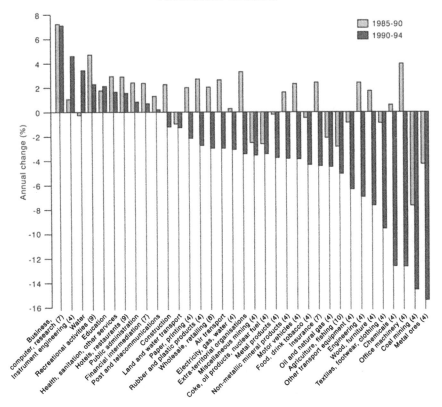

Figure 1.5 Sectoral changes in employment, per cent, E15, 1990–4 compared with 1985–90
Source: CEC, 1995b

'flexibilisation'), and recreation and tourism in Chapter 9 – as possible solutions to the employment problem in a variety of areas.

Part III brings these chapters together in terms of the outcome across the map of the EU. Sectors tend to combine in shaping distinct urban and rural outcomes (Allen, 1992) which are explored in Chapter 10 and require separate policies from the EU's 'Structural Funds'. The scope for change in EU regional, industrial and employment policies is examined in Chapter 11.

Part I asserts the importance of employment today. It continues in the next chapter by assessing the role of employment in our society and the significance of the rise of unemployment to new levels in the 1990s. In Chapter 2, jobs, joblessness and change are seen in their global context, and in Chapter 3 within the EU.

2

THE NEW UNEMPLOYMENT
IN THE GLOBAL CONTEXT

This chapter takes unemployment as the leading problem of the EU as a whole in the 1990s and asks whether the study of unemployment in other regions may help us to understand our own. After an introduction that stresses its effects, unemployment is related through the growth of productivity at work to the balance of employment and production and to the loss of secure full-time work at large. The first part of the chapter looks to the USA for *comparison*, then considers sectoral trends in other continents. Does it help our understanding of European sectoral change and flexibilisation if, for example, East Asia is showing tendencies similar to European countries?

The second part of the chapter considers 'globalisation' in terms of the *interaction* of the EU with the rest of the world, in trade competition and the effect of the EU as a leading trading bloc. This leads on to its 'internal market' and the debate about its effects on economic activity in the Union. The negative effects of growing competition from newly industrialising countries are perhaps seen most strongly in the unemployment of unskilled European workers. (The internal regional pattern of economic activity within the EU is mapped in the following chapter.)

THE SIGNIFICANCE OF UNEMPLOYMENT

Nothing has been more damaging to the social and economic fabric of this country [the UK] than the return of mass unemployment. It is both an immense social evil and a colossal economic waste. Unemployment, it scarcely needs be said, results in enormous personal hardship and misery. It destroys relationships, breeds racism and sexism and causes social disintegration. Unemployment has been the major cause of the alarming growth of inequality and poverty since 1979.

(Roger Berry, MP, 1995: 10)

Most of Berry's points now enjoy academic support, while the widening polarisation of income is internationally apparent.

22

Unemployment and health

Suggestive reports of suicides by the unemployed in the early 1980s have been followed by a mounting body of research that establishes the statistical significance of a link between unemployment and poor health. When compared with people in employment, unemployed young people and adults tend to report significantly more depression, greater general distress, lower self-esteem and more stress and anxiety (Taylor, 1991). A study of eight countries reported significant relationships between the rate of unemployment and suicide rates (Boor, 1980), while an Australian study of 401 unemployed people estimated 49 per cent to be probable psychiatric cases (Finlay-Jones and Eckhardt, 1981). Numerous researchers have found that unemployed women suffer poorer mental health, being both more committed to work and socially more isolated.

Unemployment and crime

Links are widely assumed between racist violence in, for instance, southern France or Eastern Germany and the unemployment conditions that have prevailed there. In Britain, government ministers have repeatedly argued that crime had been rising steadily even during the 1950s and 1960s when there was uninterrupted full employment. However, Dickinson (1994), taking only crime among young men, which accounts for 70 per cent of the total, found a clear association with the time series from the 1970s for unemployment in the same age group. One study of offenders found only 20 per cent of them in work at the time of the offence, while 42 per cent of prisoners aged under 21 were unemployed when they were arrested (National Children's Home, 1993). No one is suggesting that unemployment alone leads people to resort immediately to crime, but it is a powerful factor when combined with others.

Unemployment and poverty

The growth in unemployment has matched the growth in poverty attributable to unemployment very closely (Piachaud, 1991). Evidence is clearest in the UK that poverty is increasingly concentrated among the unemployed rather than, as previously, among pensioners. The highest risk of family poverty occurs in households with an unemployed head – where about 60 per cent are classified as 'poor'. Unemployment affects the income level of far more individuals than those who are directly unemployed; if a married man with three children is unemployed, this adds five individuals to the number in poverty. In addition, the real level of support provided by the Social Security system during the 1980s has been subject to growing selectivity and reduction. The Joseph Rowntree Foundation's (1995) *Inquiry into Income and Wealth* recognised that a good part of the rise in inequality was driven by structural and demographic changes that had occurred in other countries. 'But the fact that the UK suffered the biggest increase of any industrial country is largely traceable to Conservative policies' (*Financial Times*, 15 February 1995: 17).

Concentration of unemployment

Much of the rise in unemployment of the last fifteen years in the UK has taken the form of extended joblessness for certain individuals and entire households. After Pahl (1984) found that informal work and second jobs were often concentrated in already 'job-rich' households, so the reverse has also been found, that, if a household has one person unemployed, then other members are the more likely also to be without work. By 1993, six out of ten employed men had partners who were also in employment; but nearly eight out of ten unemployed men had partners who were already out of work (Gregg, 1993).

EMPLOYMENT CHANGE

Before facing 'the new unemployment', we must explore the links between employment, unemployment and the rest of the economy. Flexibilisation has made the provision of jobs more difficult to measure than before. Beyond that, however, it is important for readers, like local planners, to recognise that employment is to some extent elastic in any given area. Countries and regions with a traditionally high birth rate and shortage of work will tend to spread employment among somewhat more people than would otherwise be the case. Areas still having a shortage of labour will tend to generate fewer part-time jobs, and to leave some services less fully staffed than other areas.

In human terms, labour market behaviour and experience are very sensitive to change. On the one hand, demand for work is set both by demographic factors and the cultural changes described above by Matzner and Streeck (1991). On the other hand, the volume of employment is determined primarily by the level of activity in the various sectors of an area, including local services demanded by the size and income of the local population. These two 'sides' interact, with human adjustment occurring through *three mechanisms. One* is through migration, from areas that are short of work to others that are short of workers. The *second* is through changes in 'rates of economic activity', the proportion of people of a given age and sex who are actively seeking work. The *third* is through registering as unemployed, normally implying a readiness to take up work. Migration from areas such as North-East England or Southern Italy has always been notoriously inadequate to resolve the human problems of these areas.

The most notable feature of the last twenty years in Europe has been the response of activity rates to restructuring, both the disappearance of men from the labour market, and the increased use of women. Unemployment, however, remains the critical measure of labour market and human stress after all these mutual adjustment mechanisms have been at work. Two things can be said with certainty. Unemployment responds closely to the cyclical behaviour of national economies, with a marked timelag. There is a remarkable persistence of the same problem areas *from cycle to cycle*, connected with 'structural unemployment' from the loss of jobs mainly in the primary or secondary sectors. The arrival of new jobs can fail, however, to achieve a commensurate effect on recorded unemployment. One reason is that people may register as unemployed just to be considered for the new jobs. Or, more broadly, activity rates may rise in response to the stimulus of development. Thus, it will not be surprising if areas with traditionally low formal activity rates, for instance in Southern Europe, fail to register reductions in unemployment figures proportionate to regional development spending by the EU.

The fundamental importance of productivity

Productivity, or output per employee, remains the crucial variable both within the workplace and in national economies. It is impossible to separate the growth of productivity from the EU crisis of employment. Figure 2.1(a) shows how employment change in the EU lagged behind changes in the value of production (GDP), both in time, with a six months timelag, and in volume, in lying about two percentage points behind that of production. That is to say that a 4 per cent increase in production would yield only a 2 per

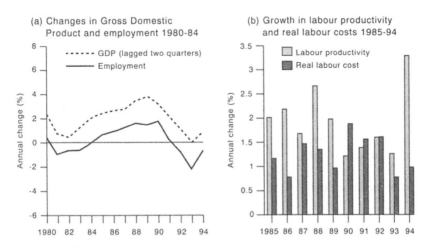

Figure 2.1 The relationship of production and employment, effected through the growth of productivity, to 1994
Source: CEC, 1995b

cent increase of employment, and that a 2 per cent increase of GDP is necessary to gain any jobs at all. The gap is due to increased labour productivity, assisted by new technology, which is shown in Figure 2.1(b) to have made a positive change in each year of the last recorded decade, but to varying degrees.

The experience of the years 1986 to 1990 was that countries that achieved a relatively high rate of output growth showed a greater than average chance of an expansion of employment. From 1990 to 1994, in the EU as a whole, GDP growth averaged 1 per cent a year, while the numbers employed *fell* by just under 1 per cent per year (CEC, 1995b). In the short term at least, a country's failure to increase productivity may help its jobs record, examples being The Netherlands and Luxembourg. I shall now consider the new levels of unemployment resulting from recession, restructuring and automation.

THE NEW UNEMPLOYMENT

The EU (E15) had no less than 17.8 million people unemployed in 1995, a rate of nearly 11 per cent, only slightly below the record of 1994. Such figures understate its adverse effects, not only because of the problems of accurate recording, but because of the limited choice of jobs available and the numbers of people who decide to leave the labour force, and therefore the unemployment figures, for a life of 'leisure in poverty'. From the positive point of view, this is an opportunity to bring more human energy back into the productive economy, so that individuals pay tax to the state rather than receiving benefits from it. This great waste of resources and loss of purchasing power almost certainly requires investment in national training programmes.

In Europe the former Communist countries have also seen a dramatic emergence of mass unemployment since 1990. National unemployment rates of over 10 per cent became common (but not universal) and threatened to rise further with continued economic restructuring towards a market economy. This follows widespread recognition throughout Europe that unemployment was a mainspring of social and ethnic conflict. Not only did it underpin the resurgence of nationalist parties and urban crime in many countries, but it was also used to accent the tensions between new international migrants and host populations within the EU countries. High unemployment in the former East Germany led to exclusion and ill-treatment of migrants from Eastern Europe and elsewhere.

The example of Germany

In East Germany (the former Communist state, re-unified with the Federal Republic in 1990) unemployment stood in 1995 at 13.4 per cent compared

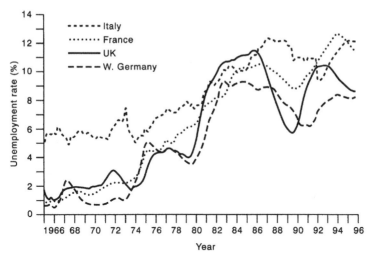

Figure 2.2 Unemployment percentage rates for the four largest countries of the EU,
standardised and seasonally adjusted
Source: *OECD Labour Force Statistics* (various dates)

with 6.6 per cent in the former West Germany. Yet the Federal Republic
consistently enjoyed rates below 1 per cent as recently as 1973 (Figure 2.2).

Even in this paragon of economic rectitude and success, rising
unemployment has become a symptom of a *Fading Miracle* (Giersch *et al.*,
1992: 197). Competition from newly-industrialising countries was seen as
one factor in the decline of manufacturing jobs, 'where it took two recessions with a net loss of more than 1.5 million jobs to achieve the required
cost adjustments by the late 1980s'. Among other factors were shortages of
capital, the German financial authorities' prioritising of inflation control
above full employment, and the increase of structural unemployment resulting from the decline of the older industries of coal, steel and shipbuilding
(Entorf *et al.*, 1990; Jones, 1994). These labour demand factors were given
greatly added significance by the expanded labour supply resulting from the
high birth rates of the mid-1960s.

The example of France

Unemployment re-emerged as the prime national issue by the 1995 presiden-
tial elections. France had experienced a greater increase in its population of
working age, one of 16.8 per cent from 1972 to 1992, which added to its
problem of student malaise and a 1995 rate of unemployment of 27.3 per
cent among 18 to 25 year olds, the highest among leading industrialised
countries.

In 1973 an annual rate of output growth of more than 6 per cent had ended, as in Germany, from the increase of world oil prices, since when lower growth and the rise of unemployment have seemed almost irreversible. The first Socialist government of President Mitterrand attempted in the early 1980s to break out of this pattern through expansionist policies, so that France temporarily avoided the worst of the recession suffered by its neighbours in 1992 (Figure 2.2). However, the expansion proved financially unviable (Halimi *et al.*,1994; Centre for Economic Policy Research, 1995) and the country was forced into accepting a higher rate of unemployment to restrain inflation. Despite structural unemployment and many relatively inflexible jobs (Gagey *et al.*, 1990), the new, 1995, government of President Chirac set a target of creating one million jobs over three years through a large subsidy to job creation schemes, and reducing taxes paid by employers taking on the long-term unemployed.

Is the UK's unemployment typical?

The UK has twice seemed a much worse case than France or Germany because of its rapid descent to high unemployment in recessions starting in 1980 and 1991 (Figure 2.2). Like many variations these arose from national exchange rates and financial policy. In the first UK recession (Townsend, 1983) it appeared that large numbers of manual factory jobs were being lost through severe but temporary circumstances, but few were restored. Fothergill and Guy (1990) later argued that recession was only a trigger that ended obsolete types of production.

The four leading European countries together (Figure 2.2) display considerable uniformity in the rise in unemployment. All exhibit a response to the events of 1973, and a sharp increase in the early 1980s. This increase was only partially reversed in the second half of the decade in Germany and the UK, before the number of unemployed rose rapidly again as the new recession started. In France and Italy, unemployment exhibited a higher level of persistence, and remained at a higher level. What is most worrying, however, is the overall long-term upward trend of the graph, which shows a deterioration unprecedented in European industrial history.

The UK composition of unemployment is the least typical. Serious as are the levels of unemployment among both young people (Roberts, 1995) and women in the UK, they are greatly exceeded in other EU countries. Table 2.1 shows the UK and Irish experience of unemployment in the 1990s to be quite untypical in gender terms, despite a near-average position with regard to age distribution and duration (the length of time for which individuals had been unemployed at the time of the 1994 Labour Force Survey). Even on comparable data, women show higher unemployment rates than men in all EU countries except the UK and the new members, Austria, Sweden and Finland. In 1995, the average unemployment rate for women in E15 stood at 12.5 per cent, compared with 9.5 for men. Six of the E12 members had

Table 2.1 Distribution of unemployment by age, sex and
duration, E12, 1994 (percentage of total)

	Women	*Aged 15–24*	*12 months and over*
Belgium	53.0	25.3	58.3
Denmark	51.8	22.7	33.0
France	52.0	25.0	37.3
Germany	50.5	13.0	30.3
Greece	57.6	39.1	50.5
Ireland	37.4	31.1	62.6
Italy	50.9	39.5	60.8
Luxembourg	50.0	33.3	33.3
The Netherlands	46.2	27.7	43.5
Portugal	52.0	34.2	42.0
Spain	48.5	33.1	52.6
United Kingdom	33.5	28.3	45.4
E12	47.7	27.6	47.7

Source: European Labour Force Survey, CEC, Eurostat 1996c

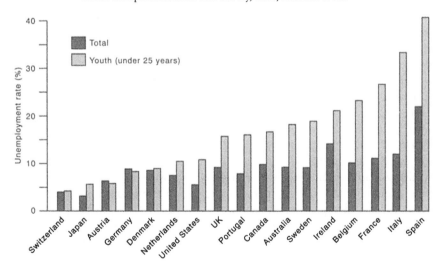

Figure 2.3 Youth unemployment rates, compared with the total for respective
industrialised countries, 1995
Source: *OECD Labour Force Statistics* (various dates)

more than half their unemployed from the ranks of women, most notably
Greece.

Youth unemployment rates above 20 per cent far exceed the all-age average in Spain, Italy, France and Belgium (Figure 2.3; full-time students not
seeking jobs are not counted as unemployed). Table 2.1 also demonstrates
the alarming position of unemployed youth between the ages of 15 and 24.

Even after discounting full-time students and trainees, no less than 27.6 per cent of the unemployed people of E12 are in this age group, a total of 4.5 million of whom no less than 1.0 million were in the single country of Italy, where they made up nearly half of the unemployed. Thus young people fail to be integrated into the labour market in many high income countries, because of educational weaknesses or minimum-wage restrictions. If the minimum wage for young people is not different from that for adults, then they will not be recruited. In particular, women have taken young people's jobs; 'the participation rate of adult females has had profound effects on youth unemployment, since adult female labour is often used as a substitute for youth (particulary male) labour. The effect is to drive down the relative wages of youth' (Anderton and Mayhew, 1994: 28–9).

Disappearing men

Even so, the men's rate would be much higher if 'discouraged workers' in the adult age bands were counted in the unemployment figures. Thus, the proportion of men who were 'economically active' in the E15 has been falling, from a figure of 83.8 per cent in 1985 and 80.0 in 1975 to 76.0 in 1994. *Adding the unemployed to the economically inactive means that working-age men without employment increased from 19.1 per cent to 31.6 per cent of the age group from 1975 to 1994.*

Virtually all member states shared in this reduction of men's activity rates over nineteen years (1975–94), a fall that continued markedly in the years 1991–4 when it more than offset the effect of rising rates for women. This most remarkable phenomenon has now afflicted prime-age men from the age of 25 to 49 as well as the younger flank (including students) and the older one (including men forced into early retirement). In Great Britain the proportion of adult men who were full-time employees has fallen below half, after excluding the retired, part-timers, the self-employed, trainees, the unemployed and unpaid family workers (*Labour Force Survey*, 1996, spring). In a new comparison of the 1981 and 1991 Censuses, the extent of GB non-employment was most marked in areas of deindustrialisation and most intense in the large conurbations (Green *et al.*, 1994).

On the other hand, it has to be said that official figures may exaggerate unemployment by including those who illegally claim benefits while working on the side, and the evidence therefore supports the proposition that precarious forms of work have been growing in several European countries (Rodgers and Rodgers, 1989). It is often said that there is a rough correlation between the size of this black economy and unemployment. In general this can be debated, but it may explain how Spain has survived an average unemployment rate of 18 per cent over the last decade without huge social unrest. 'Many of Spain's "unemployed" probably do work' (*The Economist*, 12 February 1994: 25).

THE GROWTH OF INSECURE WORK

In addition to unemployment and non-employment we must add the newly recognised field of insecure (flexibilised) employment, which appears for the first time to include sizeable numbers of the middle classes. It was not entirely fanciful for writers to create transatlantic titles on *The Culture of Anxiety: The Middle Class Crisis* (Symonds, 1994), *The End of Work* (Rifkin, 1996) or *Jobshift* (Bridges, 1995):

> Only a minority of jobs in Britain now fits our idea of a 'normal' job or 'real' job. We are only just beginning to understand the implications of this. In the last 15 years the labour market in the UK has undergone more radical changes than occurred in the previous half century.
>
> (Brewster, 1995: 2)

This threatens the design of the welfare state, based on lifelong work except for a few periods of unemployment and a small number of disabled. We have already noted that 31.6 per cent of adult men were *out of employment* (41.3 per cent of both sexes) in E15 data of 1994 (CEC, 1995b). What is new is the recognition of a further *group of 'flexible' or 'insecure' staff*. In defining his '30/30/40 society', Hutton (1995) included in the second 30 per cent those with insecure jobs, poorly protected and lacking fringe benefits; the remaining 40 being the *full-time employees and self-employed* who have held their jobs for more than two years. E15 data for 1994 show 4.8 per cent of men to be working part time, 10.1 on temporary contracts, and 18.9 to be self-employed (these groups overlap but would sum to 33.8 per cent, 41.4 of both sexes). Although we cannot be precise, there is little difficulty then in applying Hutton's new concept of a '30/30/40' flexibilised labour force to the EU as a whole, as read from its average (although Southern Europe, as before, had never fully adopted the formal structures of an industrialised society in any case).

The middle classes had previously condemned the unemployed as work-shy. What is new, as from the US and European recessions of the 1990s, has been a recognition that no amount of geographical mobility will guarantee a job for life for graduate staff, as recognised here by a (black) trades unionist:

> Many of those in between [the rich and the poor] have been trapped in a net of anxiety that comes from having far less security. Job insecurity has seeped deeper and deeper into all sections of the workforce - from the factory floor to the bank manager's office. Should ill fortune strike, the social security net seems increasingly threadbare.
>
> (W. Morris, General Secretary, TGWU, 1995: 3)

Redundancies of the 1980s meant that there was no surplus labour left to cut in firms in the 1990s except among white collar staff. Job losses among middle managers were at first cost-driven and piecemeal, but led corporations on both sides of the Atlantic to theorise improvements in efficiency by stripping out bureaucratic functions:

> In other words, what had begun as a response to recession, turned into a fundamental re-evaluation of the nature of the corporation. Downsizing and delayering are the new watchwords and it has meant large numbers of previously secure middle class, middle management jobs being consigned to corporate history. . . . British Gas, BP, Shell, ICI, retailers like Tesco, Sainsbury and WHSmith and all the clearing banks have been getting rid of managers at all levels.
>
> (Symonds, 1994: 14)

Some qualifications

But is this an end to full-time work? There is some danger of generalising from the early 1990s recession. In the USA, some 'delayered' managers were being sucked back into full-time staff jobs, as companies started to grow again (*Financial Times*, 3 April 1995). Others argued that the talk about the end of careers and jobs came rather subjectively from media people, lecturers and consultants who had themselves lost secure employment, and that all the evidence of temporary contracts was still strongly linked to the economic cycle. In the whole of E15, full-timers still made up 84.7 per cent of total employment, and employees represented 84.9 per cent (CEC, 1995b). Redundancy in Great Britain had fallen back to 213,000 in the autumn quarter of 1995 compared with the quarterly peak of 344,000 in the winter of 1992, with higher proportions finding work in the same quarter.

The initial evidence still suggests a *direction of change* moving consistently away from conventional full-time work. In a review extending across western Europe, Rodgers and Rodgers (1989: 9) argued that 'The evidence therefore suggests that precarious forms of work have been growing in several European countries, although the growth is far from uniform across countries and does not affect all atypical forms of work.'

Although then we are far from talking about 'the end of work', we can say that employment growth in western Europe has been very limited in quantity and quality. In fact it has largely been confined to five years out of the last twenty, i.e. 1985–90. Numbers of men workers in particular have ceased to grow in nearly all countries, and fell by 5.4 per cent in the recession years of 1991–4. Wages and salaries have been declining relatively as a source of income. In the UK they represented 68 per cent of household income as recently as 1971, while this share had fallen to 57 per cent by 1992 (*Social Trends*, 1994), mainly between the years 1979 and 1987 when recession caused an increased dependence on social security payments, rents, dividends and interest. The significance of all these changes for individual geographical areas is that their sources of income are more mixed and uncertain than before.

CAUSES OF HIGH UNEMPLOYMENT, COMPARED WITH NORTH AMERICA

'Europe as a whole is an unemployment blackspot' (*The Economist*, 12 February 1994: 80). The major EU countries collectively distinguish themselves as a group from the rest of the industrialised world; indeed their worse performance is said by right-wing writers to give away their policy mistakes. I will now consider more data for the EU as a whole, and the value of more comparisons with North America. By mid-1996, the EU had 11.2 per cent of workpeople out of work, double the figure of 5.4 per cent for the USA. The recession of the early 1990s merely confirmed an alarming underlying trend in Europe and a further divergence from the USA.

Only twenty years earlier the position was the reverse; Figure 2.4a shows that E15 had only half the unemployment rate of the USA. The massive rise in Europe compared with the US is most apparent after 1983 when the USA staged a much more successful recovery from recession. While the US jobless rate tended to decline during the 1980s and 1990s, Europe's dole queues got relatively worse through the cycle, with no permanent benefit from the 1985–90 economic recovery, to a peak of 11.4 per cent in 1994. The principal casualties have been from the creation of more long-term unemployed, as no less than 47.7 per cent of the unemployed in the European Labour Force Survey, 1994 (CEC, Eurostat, 1996c) had been out of work for twelve months or more.

Figure 2.4 The deteriorating employment trend of E15, 1972–94; (a) out of work but seeking it ('unemployment rates'), (b) in work ('employment rates')
Note: includes the newest members of 1995 (Austria, Finland and Sweden), compared with the USA and Japan
Source: CEC, 1995b; CEC, Eurostat, 1996c

The USA has far more turnover in its labour market, and a key point is whether this simply constitutes a symptom of healthy growth or is itself a cause of lower unemployment in constituting a 'more efficient labour market'. In the USA, the chances of becoming unemployed in any one year are several times higher than in the main EU countries, but the chances of finding a job are even higher, resulting in less than 10 per cent remaining unemployed for more than one year. 'Higher unemployment in the EC countries is not so much the result of people frequently losing jobs, it is more to do with a lower probability of finding a job having once become unemployed' (Anderton and Mayhew, 1994: 18–19).

The rise in European unemployment over the past decade is sometimes blamed on demography. Changes in total unemployment reflect the difference between the growth in the labour force and the actual change in employment. The faster the growth in the labour force, the faster employment must grow simply to hold unemployment steady. Europe's adult population did grow over twenty years (Table 2.2), but at an appreciably slower pace than in Japan and North America, where a faster growth still of the economically active (including in this case men) caused their labour force to increase almost four times more rapidly than in Europe, where *employment rates*, the proportion of adults in employment, actually declined (Figure 2.4b). At the same time, the growth of jobs in Europe, those 'in employment', failed even to match its modest labour force growth, resulting in almost the quadrupling of European unemployment.

Thus, we can see that Europe's big problem has been its overall net failure to create jobs (even though demography has contributed to the problem at certain times and places, such as Southern Italy). Sometimes the distinctive

Table 2.2 Twenty years of divergence between the industrialised continents

	European Union millions			Change 1972–1992 (%)		
	1972	1992	Change 1972–1992	EU	Japan	North America
Population, 15–64	194.1	213.0	+18.9	+9.7	+19.0	+28.2
Economically active	127.1	143.3	+16.1	+12.7	+26.5	+45.6
In employment, incl.	123.4	129.3	+5.9	+4.8	+25.6	+42.3
Agriculture	15.1	6.9	−8.2	−56.1	−45.6	−6.2
Industry	49.3	39.8	−9.5	−19.3	+19.7	+8.3
Services	55.2	80.6	+25.4	+46.1	+51.3	+65.2
Unemployed, nos.	3.7	14.0	+10.3	+274.0	+91.9	+101.3
Unemployed, %	2.9	9.8	+6.9	+237.9	+57.1	+37.5

Source: OECD, 1994a

attribute of North America is seen as the growth of service sector jobs. However, Table 2.2 also reveals that services in the EU were almost as effective in percentage terms in providing jobs as was the case in North America. As emphasised by Glyn (1995), the difference between the two continents lay in a faster loss of jobs on the land and in industry. In the North America of the 1970s, agricultural employment had long been mechanised and stood at a low level, while the exodus from the countryside was in full flow still in areas such as Spain. (In North America, industrial employment actually increased in absolute terms, in 1972–92.)

The constant underlying factor in these equations is again that of the growth of productivity. Thus if EU output was increasing by 2.1 per cent per annum from 1975 to 1988, and labour productivity, output per person, was increasing by 2.0 per cent per year, then it is unsurprising that employment grew at a very low average rate, in fact one of 0.1 per cent per year. This was lower than in earlier decades because of the later reduced growth of GDP and because Europe was still catching up on North American levels of productivity. Despite a huge increase in computing over the past decade, unemployment in the US is little higher than twenty years earlier; in western Europe, where the investment has been less, unemployment has been higher. This is consistent with the OECD's (1994b) *Jobs Study*, which concluded that information technology (IT) played a moderately beneficial role for employment, or Glyn's (1995) view that it was absolutely wrong to blame technology for destroying jobs.

Unfortunately, there may be prolonged lags between job losses and the creation of new jobs, evident in the wider European gap between the loss of agricultural and industrial jobs and the growth of services. The OECD (1994b) argue the now orthodox 'neo-liberal' view that the process of change is facilitated not only by high levels of overall growth, but by flexible markets for both labour and products. Governments can therefore help by providing education and training, and by removing obstacles to the free market in goods and services.

It is here that neo-liberals bring in the USA again. While the US is seen to benefit from a 'deregulated' labour market and high turnover of the unemployed, Europe is seen to suffer from measures of job protection provided in better days by more paternalistic regimes. Europe, on this view, is making it harder for potential employers to take workers on, and thus they do it less. The 'efficiency' of the less regulated US jobs market has been compared with 'Eurosclerosis' (Giersch, 1985), with the following questions at issue:

1 Are wages in the EU too high, supported by rigid industrial relations?
2 Have minimum wages had negative effects?
3 Have non-wage labour costs, such as employers' contributions to National Insurance, had adverse effects?

4 Have unemployment benefits been too high to encourage workers to take work?
5 Are workers too well protected from dismissal, for instance through a requirement for compulsory redundancy payments?

A right-wing analyst of these issues would take it as read that their relative freedom from regulation in the USA was the very thing that prompted so much job growth.

Where there is agreement is that *the level of unemployment which is consistent with an absence of rising inflation has risen*. On this argument, governments have greatly reduced their practical commitment to full employment in order to control inflation in the economy. Thus, a government might decide its financial policies so as to deliver an unemployment rate of, say, 7 per cent, when it had hitherto achieved the same level of inflation with, say, 3 per cent. When governments have been bound together under the European Monetary System, there has been a deflationary bias at work that has certainly been among the more immediate causes of unemployment.

We do not know why the sustainable level of unemployment has increased, but the 'natural unemployment rate' certainly underpins the persistence of high unemployment rates, once they have been established, through, for example, structural change. However, a major question remaining for the rest of this chapter is that market prices have turned against unskilled labour, taking the form of low wages in the USA and the form of unemployment in Europe. In the US, 'the sizeable reductions in pay for the less skilled have not been sufficient to maintain their employment; have impoverished them and their families; and arguably contributed to the decision of many of them to engage in crime' (Freeman, 1995: 70).

Nevertheless, there are recent suggestions that the natural rate of unemployment has begun to decline (Bowley, 1996). Most mainstream writers give out an uncertain sound on the causes of European unemployment, and most geographers will see a need for labour market regulation (e.g. Peck, 1994). The International Labour Organisation (ILO) (1995: 156) concludes that it is 'far from clear that labour market rigidities are the major cause of high unemployment'. All high income economies have shown difficulties in their labour markets, and labour market performance has deteriorated compared with the growth environment that prevailed before 1970.

SIMILARITIES OF GLOBAL TRENDS?

The ILO (1995) also claimed that on a world basis, 30 per cent of the world's workforce can be estimated as unemployed or underemployed. The sectoral shift towards service and women's employment is now common in most parts of the low income countries. The study of patterns outside Europe will reinforce our impressions of many trends that are occurring in our own societies. Full-time wage labour for men can be seen as an aberration of the West from the late eighteenth to the late twentieth centuries (Pahl, 1984). If the 'flexibilisation' of work is providing Western economies with a departure from 'conventional' work, then this may represent a return to widespread informal work, which used to flourish in the West, and is expanding in low income countries.

Sectoral shift

ILO production data shows 'convergence' between industrial and low income countries (Figure 2.5). The poor countries have followed the industrial ones in expanding industrial and services output at the relative expense of agriculture. The 'globalisation' of industry has actually given the poor countries virtually as great a dependence on manufacturing industry (20.7

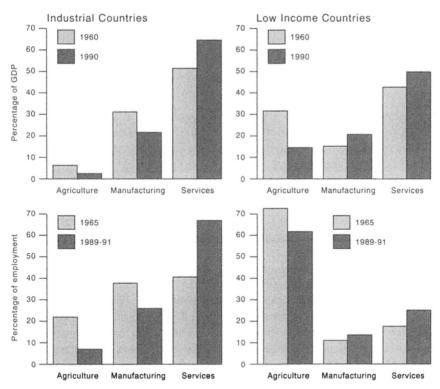

Figure 2.5 Convergence in the structure of production (GDP) and employment between industrial and low income countries, 1960s–1990s
Source: United Nations: *Statistical Database* (New York, 1993) cited in ILO: *World Employment 1995. An ILO Report* (Geneva, 1995), p. 30. © International Labour Organization

per cent in 1990 and rising) as the so-called 'industrialised' economies (21.5 and falling), with (remarkably) the majority of poorer countries' exports now comprising manufactures. High growth was achieved in the favourable export performance of East Asia through manufacturing exports, notably from South Korea, Taiwan, Hong Kong and Singapore, boosted by significant inflows of foreign direct investment (FDI) in several of these countries. In this sense, successful response to the opportunities created by globalisation was a key explanation of their superior economic performance.

Figure 2.5 also shows structures of employment and, by comparison, thus gives an indication of differences in productivity between sectors. Most agricultural employment is now in low income countries, where productivity is very low. Poor countries' growth in the service economy is a less familiar feature in the literature, beyond the mushrooming of the urban informal sector in major cities.

However, a moment's reflection will remind us of the early importance in the European-dominated globe of the office complexes of Buenos Aires, Argentina or Johannesburg, South Africa, both recognised today as at least satellites of the 'World City' system. Transnational corporations (TNCs) in the service sector, whether in finance or the hotel trade, follow their global manufacturing customers in establishing themselves in a full network of these cities, accompanied now by the extension of world tourism to many tropical destinations.

> Beginning in the 1970s, the sleeping giant of the world-economy awoke and began to stalk the globe. Until then, service industries had tended to be domestically bound. . . . By the 1980s service industries were growing faster than any other sector of the world economy.
>
> (Thrift, 1989: 32)

Many of the fruits fall to the richest countries, as four-fifths of world trade in services comes from only twenty countries. However, services already represent more than 60 per cent of the economies of the four fastest growing countries of East Asia, which have established their own hotel, airline and financial groups and intend to build international sales across the whole area. 'The growth in international trade and investment in services offers unique opportunities for employment generation in developing countries with an educated labour force' (ILO, 1995: 47). An example is that of the Indian software industry which is achieving sales in the industrialised world.

It is more difficult to translate these types of data into employment estimates. Daniels (1993) agrees that only a few special cases such as Singapore or Bahrain had levels of service employment approaching those of high income countries, but all his examples showed service employment increasing its share of total jobs. The EU's increase of service jobs is occurring against the background of very general growth of service jobs elsewhere, linked with measurable growth in women's share of the labour market.

The informal sector

In the early 1970s, at least a part of the additional employment in commerce and other services occurred through work-spreading and income-sharing (Bhalla, 1973). The informal sector of poor countries may have more to tell us than we had realised. 'Informal' work in a poor country varies from cottage industries in rural areas to shoe-shine boys and more organised manufacturing in cities. By definition, employment is usually not governed by contract or other legal conditions, and participants are ineligible for social security benefits, legal protection or trades union membership. Work requires few educational qualifications and is labour intensive, with flexible working hours, often part-time or falling outside the normal working day, with many participants having multiple part-time jobs, often self-employed or working as a household.

In the 1970s, attention focused on employment-intensive strategies in the informal sector which were seen as more appropriate to situations of extreme labour surplus. The ILO, rather than viewing the mass of people who were not in formal employment as a problem, saw them as a resource, a view that resurfaced in the 1980s. It was the informal sector that absorbed additional labour during the 1980s while the modern, formal sector was reducing its total workforce and levels of wages, as in Latin America, although it was realised that their prospects depended entirely on the formal sector. By the mid-1990s (ILO, 1995: 75), 'countries are likely to do better on the employment front if they adopt policies that respond positively to globalization', provided that it is recognised that not all countries have a comparative advantage in manufacturing and that agriculture is supported.

There must be at least a suspicion that the growth of non-conventional 'flexibilised' work in areas such as Southern Italy and the UK may show increasing parallels with the poor countries' informal sector. This is the point, however, at which we move on from *parallels and similarities* between First and Third Worlds to looking at competitive global *interaction and competition*.

GLOBAL INTERACTION

A prominent business journal wrote of

> The *paralysed economies*, chiefly poor ones in Africa and elsewhere that are going backwards or at best standing still. The *progressing economies*, those in Asia, Eastern Europe, Latin America or even the richer world that are actually moving forwards in a sustainable way. And finally the *paranoid economies*, those in America, Western Europe and Japan that are terrified by the competition from the progressives.
>
> (*The Economist*, 6 May 1995: 24, *emphasis added*)

Is such paranoia justified in the EU? That is the question for this section, which will explore the overall employment issues of the EU in the geographical context of 'globalisation'. Should Western Europe fear anything that Japan or the USA does not? In taking a more geographical approach than in the last section, it is possible here to link the development of industrialisation in other continents with events within the EU area, for instance the relationship between successive phases of Asian industrialisation and the levels of EU unemployment in textiles, which will relate to producing areas of the member states in the next chapter. We now need therefore to see all the employment issues of the EU more geographically, as part of an international pattern and chain of causation.

> In the past unemployment in Europe has often been blamed on 'cheap foreign imports'. More broadly, Europe as a whole now faces demotion in terms of its economic power. For instance, a leaked UK government document argued that 'if recent growth trends continue', China would become the 'single largest [economic] nation' in 2015. It predicts India in fourth place, Brazil sixth and Indonesia seventh. While the US, Japan and Germany would remain in the top seven, France, the UK and Italy would slip into the second division.
>
> (*Financial Times*, 7 June 1996: 18)

Globalisation

Economic change in our individual countries is directly subject to global change. The simultaneous production and consumption of the same products and images is found in every corner of the globe. The economic, social, cultural and political effects can be summarised under the term 'globalisation', described by Amin and Thrift (1994: 8) as representing 'above all, a greater tying in and subjugation of localities (cities and regions) to the global forces'. This by no means represents the end of local determination of events. We cannot read off all local outcomes from world trends, nor are TNCs incapable of interacting with and being influenced by the localities in which they operate. 'Globalisation' has not yet reached the rural areas of all poor countries. Nonetheless, a transition can be seen, starting from the early 1970s and most obviously in the economic sphere, in which international interaction accelerated to the point of creating a qualitatively different world. Above all, this is because in every decade and almost every year since the Second World War, world manufacturing trade has increased at a faster percentage rate than that of production.

Much of this has cultural as well as social and political implications. The pressures of international business and travel on the cultural integrity, especially of non-English speaking areas of the world, are considerable, with McDonald's hamburgers being all the rave in Moscow and EuroDisney having provided a large (though still disappointing) number of jobs with their development at Marne-la-Vallée near Paris. The fundamental effects of the international economy on social geography are profound. For example, the growth of white collar jobs in the new business services of leading 'world cities' has led to profound debate about polarisation of income and social structure of fundamental significance in the division of society (Hamnett, 1994; Sassen, 1991).

The context for global corporations

What is new about 1990s globalisation? Much of the world was receiving European and US investment in primary production under their imperial control in 1914; it was in that sense 'globalised'. The convulsions of the 1930s, when trade dried up and killed democracy in most of Europe, were the result of mishandling of the international financial system. This was, however, improved sufficiently after 1945 to generate Hobsbawm's (1994) 'Golden Age' of post-war economic growth, or the 'Long Boom' of Marxist analysts, involving high rates of growth of trade and industrial output until about 1974. This was accompanied by structural change in the heyday of 'Fordism', the industrialisation of Southern Europe and the movement of large proportions of the rapidly increasing populations of poorer countries to the cities, notably in the industrialisation of Brazil and Mexico.

The period since the 1960s, however, has been substantially different. On the one hand, deep recessions caused growth to falter in the Western economy. The geographer Ahnstrom (1990) accepts that the change in the EU from economic growth to *'stagnation'* from 1974 was triggered by the international financial crisis of that year; reduced public investment was connected with reduced demand for consumer goods. On the other hand:

> in the 1960s FDI grew at twice the rate of Gross National Product (GNP) in OECD economies; in the 1980s it was growing four times as fast. . . . In terms of direct investment alone therefore there is a dramatic shift towards a new level of integration in the world economy.
>
> (Howells and Wood, 1993: 4).

Much analysis of FDI focused on a 'new international division of labour', in which TNCs of the rich countries established branches for world market production in poorer countries. TNCs could now separate out different phases of production between, say, R&D in rich countries and routine production in poor, aided by changes in the technology of production and communication (including use of containers and air freight) and by organisational changes in the international capital market and in countries' taxation arrangements. Middle income countries such as Mexico, Tunisia or Malaysia established 'export processing zones' in which their plentiful supply of available labour could be tapped and controlled in manufacturing establishments that paid no customs duties on imported materials used in manufactured exports.

Although the transfer of investment in individual corporations is potentially very serious for the EU, I consider that this view of 'globalisation' has been somewhat overwritten, and is already seriously out of date. According to the United Nations Conference on Trade and Development (UNCTAD, 1994), TNCs employ only 12 million workers in poor countries (2 per cent of the total, including services as well as factories), compared with 61 million in rich countries. There are, however, two new features. TNCs from rich countries are placing a fresh kind of investment in poor countries involving greater amounts of skilled work, R&D and independence for establishments there. On the other hand, the TNCs of East Asia have followed Japan in providing a return flow of investment to the EU and USA; in 1996, UK regions were attracting a very significant proportion of Korean factory investments to the EU. Looking at the overall picture, however, there are still very few genuinely worldwide TNCs, and the term 'globalisation' can be a rather exaggerated conceit of industrial geographers, unaware of how much of Tropical Africa, for example, is dropping out of the arena of world progress.

The fall of the rich industrial countries?

The *overall* trading power of newly-industrialising countries still gives great cause for concern as they continue the process of 'catching up'. Greater economic power for other continents must at least change Europe's mode of interaction with them. Seen from the viewpoint of UK welfare, 'technological change is fuelling growth in developing economies at a scarcely believable pace ... the result of these fast growth rates among newly-industrialising countries is a new international geography of economic strength' (Report of the Commission on Social Justice, 1994: 67).

It may come as a surprise that China accounted for as much as 30 per cent of world manufacturing output as recently as 1830, while what are now the 'middle/low income countries' still had an overall majority of industry, mainly handicraft. Figure 2.6 also shows how the 'rich industrial countries' attained their maximum share of world output as early as the 1920s, prior to their gradual decline to just above 60 per cent, partly connected with the success of the former Soviet Union and Eastern Europe in raising their visible share (on the graph) from 1928 to 1963, but more recently with the cumulative acceleration in lower income countries as from the 1960s. The rich industrial economies' dominance over the world economy is already less than is generally realised; revisions by the International Monetary Fund based on the actual purchasing power of currencies, rather than their official exchange rates, put the rich countries' share at just 56 per cent of world output. Within the rich countries, Japan increasingly became a threat to both the USA and Western Europe.

42

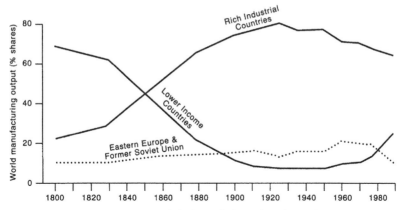

Figure 2.6 The economic rise and decline of the 'rich industrial countries' in the
nineteenth and twentieth centuries
Source: *The Economist* 1 October 1994 (after Paul Bairoch)

West Germany, calculated without the East, retained a higher income per capita than Japan in 1992. This was a close-run thing:

> Instead of being a challenger to technologically more advanced countries like the US, West Germany had to face challenges from below, namely the rise of Japan and various developing countries whose emergence had been observable as early as the 1960s but which had been much less noticeable at that time.
>
> (Giersch *et al.,* 1992: 222)

In all, imports from beyond Europe and the USA increased from 12.9 per cent of the total in 1965 to 22.1 by 1988, the change being attributable first, to Japan, second, to newly-industrialising countries in general, and third, to the 'Four Little Tigers'.

The 1980s pattern of change

The 1980s showed great divergence between different countries, as Latin America was held back by massive international debts, and the progress of Asia was actually joined by India and China among countries with the highest growth rates in production (Figure 2.7). In some countries, growth rates exceeding 8 per cent per annum were achieved over the period 1980–93, notably in South Korea, China and Thailand. Even across whole sub-continents, rates of growth of GDP of 7.8 per cent were found from 1980 to 1993 in East Asia and the Pacific, and 5.2 in South Asia, compared with 2.9 in the high income economies as a whole, and 2.5 in the UK.

These trends of themselves went a long way to prompt the idea of a 'Triad' of continental economic power (Ohmae, 1985), centred on Japan,

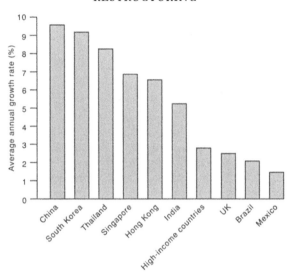

Figure 2.7 Leading countries for annual growth in production (GDP), 1980–93
Source: World Bank, 1995

North America and Western Europe, with each area exchanging FDI with the others, and extending it to its respective 'tributary' sections of the globe. Within this pattern, trade was becoming increasingly bitter between the US and Japan, the US was becoming much more a recipient of inward foreign investment, but 'Europe was undoubtedly weaker than its two major rivals' (Dicken, 1990), notable for overcapacity in the motor vehicle industry and for a much better position, for instance, in pharmaceuticals than in electronics.

If Western Europe or even Germany were paranoid about overseas imports of manufactured goods, then this was really applicable mainly to East Asia. If we allow for the greater growth of trade than of output across the industrialised world, *trade with poorer countries, in terms of trade with poorer countries overall, was actually growing less than was trade between rich industrial countries.* So we arrive at a rather remarkable conclusion:

> Although many developing countries have captured markets for their consumer goods industries in the industrially advanced countries since the beginning of the 1960s, it is unlikely that imports from the Third World countries en bloc represent a major threat to capital accumulation and economic growth in Western Europe.
>
> (Ahnstrom, 1990: 134)

Remarkably, other authorities agree that expansion outside the industrialised world, even a massive growth of the Chinese economy, can only help Europe in the longer term, through providing markets for its more sophisticated exports. However, most areas of the world, even that of East Asia, are forming themselves into defensive continental trading blocs not unlike the EU.

THE MERITS OF CONTINENTAL TRADING BLOCS

In the pursuit of 'making a living in Europe', it may seem paradoxical in an era of successful free trade to find the EU attempting to protect itself with a common albeit low external tariff and other restrictions. However, much of the identity of the present-day EU results from the need for European countries to ally together in facing the world, perhaps more the competition of the USA and Japan, we find, than the poor countries as such. The regional trading bloc is a common feature now well imitated around the world, which might be seen as a 'second best' or step on the way to universal free trade (Gibb and Michalak, 1994).

In the industrialised world of the 1930s, countries met the severe reduction of world trade through a 'beggar-my-neighbour' policy of erecting high tariffs against outside imports. However, this only increased their own unemployment to higher levels. The lesson applied by GATT was to reduce international tariffs from an average level of 40 per cent in 1948, gradually to one of about 5 per cent by the end of the 1980s. GATT comprised a multilateral treaty implemented through a series of 'rounds' to improve the rules of trade. Some three-quarters of international trade lies between industrialised countries, containing 15 per cent of the world's population. The more recent progress of GATT would appear to favour poorer countries, in reducing subsidised overproduction of food in Europe, and reducing rich-country protection against textiles imports, through the 'multi-fibre agreement'. However, opinion is that GATT has in practice worked in favour of the TNCs and of high to medium income countries.

Notwithstanding the theoretical merits of totally free trade, it is felt that it carries particular dangers in real historical situations. The idea of regional trading blocs was strongly revived and extended in the 1980s and 1990s, building on the precedent of Europe's Treaty of Rome of 1957. The individual nation state, as in Europe, is acknowledged as too small to achieve many of the economies of scale of modern technology; but many countries wish to stop short of full exposure to global free trade, and seek that in regional trading blocs. The EU has acted as an example to the rest of the world in building on earlier sectoral co-operation gradually to remove internal quotas and tariffs and to adopt a common external customs tariff.

It is, however, in part due to Europe's relative economic decline *vis-à-vis* North America and South and East Asia that the drive for a Single European Market between the E12, now E15, has been pursued strongly.

It is quite clear that the prospect of a unified market of some 320 million consumers, set within a single external Community customs boundary, has created enormous interest and unease amongst Europe's trading partners and among foreign firms seeking to sell in the Community.

(Dicken, 1990: 47)

It was partly as a result of inter-continental competition that the USA sought the agreement of its neighbours Canada and Mexico in the establishment of the North American Free Trade Area (NAFTA), operational from the beginning of 1995. Significantly for the UK, job losses in Canada were expected to continue to be focused in a narrow band of the labour force, notably in low wage, labour-intensive manufacturing, particularly affecting women in textiles and clothing (England, 1994). NAFTA did not prevent the USA making overtures to the other regional groups in Latin America, nor from establishing 'Asia-Pacific Economic Co-operation' (APEC), including China, Japan and sixteen other members including the Four Little Tigers. Eastern European countries predominantly wish to join the EU, adding to the complexity of its external relations, already well stretched by commitments in Africa and to other former colonies of the European countries.

The internal market

If the external effect of the EU has been to safeguard some jobs from external competition, it has also increased internal activity and trade, creating *at least some* greater growth than would otherwise have occurred. The logic of the European Union ran deeper than other continental customs unions in envisaging, successively, a Common Market, an economic union, and a political union. The concept of the Common Market was to allow free movement of land, labour, capital and services between member countries, so that factors of production were equalised between constituent areas. Beyond that, the concept of economic union involved the harmonisation of economic policies and the development of supra-national institutions. For example, environmental controls over pollution and the impact of major projects were standardised, so that a country with lax laws did not automatically attract all the investment in a particular industry, once barriers to trade had been dismantled. The concept of political and monetary union is one that generates most remaining controversy in Britain, and could lead to much more powerful supra-national political institutions.

While holding to about a fifth of world trade with its external transactions, the EU greatly increased the internal trade between its member states, notably between 1957 and 1973 with the dismantling of internal tariffs, and between 1985 and 1992, during the implementation of the Single Market, when many of the remaining (non-tariff) barriers were dismantled. The effect of these changes, including gains from economies of scale and intensified business competition, were *expected* to yield an addition of 4.3 to 6.4 per cent to overall output of the EU (Cecchini, 1988). 'The cumulative effect on GDP may have reached 3.5% by 1990 (1.5% on employment), remaining at that level during 1991 and 1992; i.e. it is still present underneath the present recession' (Reichenbach *et al.*, 1994). Others, for example Sadler (1992), argued that the impact of cost reduction and increased demand, for example on the motor industry, had largely been overstated by the EU, while Lindley and Wilson (1992) foresaw a range of possible employment outcomes in Britain, between 4 per cent *gain or loss* for the period 1989–2000, with some of the greatest possible variation in chemicals and banking.

THE GLOBAL EFFECT ON EUROPE'S UNSKILLED

We have already accepted that the effect of manufacturing trade competition from poorer countries may have been overstated and may be substantially less than from fellow industrialised countries. However, we have also noted the expected disproportionate effect of NAFTA on more labour-intensive industries in Canada, and the scale of German adjustment over time to poorer countries' competition in textiles. Bringing the themes of this chapter together, it may be that unemployment among *unskilled* workers is more heavily linked to the overseas trading position.

Poor countries will certainly remain a scapegoat for industrial change. As emerging countries' exports, boosted by GATT's Uruguay round, continue to grow, so will the resentment of many people in today's rich nations. Europe and sometimes the USA is awash with such pessimists, with French critics arguing that free trade with lower income countries will lead to mass unemployment and huge wage inequalities. Of course, cheap imports date back as an issue to the inter-war effects of the Indian cotton industry on Lancashire. What is new this time is the sheer weight of new competition from Third World labour. With Chinese and Indian workers willing to accept an average wage of little more than 50 cents an hour, compared with average labour costs of around $18 an hour in rich countries (Figure 2.8), then it is claimed that fierce competition will steal our jobs.

The World Bank (1995) in its Report on *Workers in an Integrating World* notes that there have indeed been more exports of labour-intensive products and services from low income to high income countries. Other economists go so far as to say that the increase was the main cause of substantial change in labour markets. Wood (1994) notes that the growth of wage inequality, especially notable in the high income countries of the UK and USA, occurred at precisely the same time as the renewed growth of imports from poorer countries; indeed the latter will explain almost the whole of the

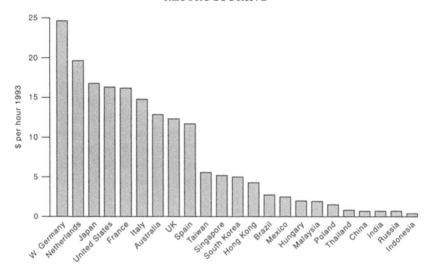

Figure 2.8 The significance of cheap labour in low income countries; costs per hour
including non-wage costs in manufacturing
Source: *The Economist* 1 October 1994 (after Morgan Stanley)

collapse in demand for unskilled labour. Minford (1995) sees the same effect
as the clue to growing wage disparities in several countries.

> The demand for unskilled workers has been falling in the OECD coun-
> tries for many years and, as a consequence of the expansion of higher
> and further education, so has the supply. But, since the beginning of
> the 1980s, there is some evidence that demand has been falling faster.
>   (Nickell and Bell, 1995: 40)

There is substantial international evidence of the worsening position of the
less educated. Unless they are well qualified, job-seekers are faced with predominantly part-time, insecure work at pay rates that are declining relative to
the average and usually provide no clear route to a better job.

Certainly theory will support the view that, in the course of intensifying
exchanges with less advanced and hence comparatively labour-abundant
regions, it is those sectors of industrialised countries that make the least
efficient use of labour that have to bear the burden of adjustment. If trade is
opened between two areas, both sides gain, but not everyone inside each area
need do so.

Detailed studies have shown that what were originally low wage countries
exerted considerable pressure on industries with a high share of low skill
employment and on industries concentrated in relatively backward regions
of West Germany (Giersch *et al.,* 1992). In other words the challenge was
most pronounced in those branches exhibiting features similar to those of
the poorer countries themselves. These led to some pressures for the adop-

48

tion of protective tariffs, and analysis showed that the sectoral profile of German state assistance revealed a significant bias in favour of declining industries, often identical with those under competitive pressure from poorer countries.

However, there must be doubts about Wood's and Minford's views. Poor countries are responsible for only 12 per cent still of EU imports, and in turn the great majority of our unskilled are now employed in services. Professional economists have thus been unable to offer a conclusion (Flanders and Wolf, 1995). Many accept that trade plays some role in labour market trends in Europe.

CONCLUSION

Whatever the outcome of the argument, it would not prompt many economists to favour a return to protectionism. As we said earlier, many arguments point to great benefits accruing from increased trade with, say, India and China, especially in finding export opportunities for European goods. There is little evidence that use of flexible technologies may in future allow TNCs to bring back automated production to Europe (Echeverri-Carroll, 1994).

The Centre for Economic Policy Research (1995) concludes that the rise in European unemployment is not principally due to *any one* cause, whether competition from poorer countries, rapid technological change, the generosity of welfare provision, or labour market rigidities. Specific reforms, they say, would relieve unemployment, but there is not necessarily a democratic majority likely to support such reforms. Glyn (1995) concluded that while the higher unemployment rates needed to prevent accelerating inflation were still not understood, there was little doubt about *their* disproportionate effects on the unemployed rate of the unskilled.

The precise relative importance of different causes in the rise and persistence of high unemployment in Europe remains a matter of debate. Even so, the depth of the early 1990s recession and the problem of social exclusion have given fresh impetus to the political significance of unemployment throughout Europe, and to the weight given to it in the European Commission. The 'end of conventional work' is not as close as the polemicists have said; nonetheless, there are reasons for looking yet more positively at the use of leisure time (Sayers, 1988).

SUMMARY

- Unemployment has increased in importance in the EU, not only because of its persistence at near record levels, but also because of its sensitive social effects on health, crime and poverty.
- While recorded rates of unemployment in most countries (not the UK) are worst among women and youth, the effects of early retirement and

other factors now mean that nearly a third of EU men of working age are not working.

- There is a lower probability of an unemployed person finding work than in the USA, but the poor performance of the EU is due to the greater long-term loss of farm and industrial jobs.
- The EU growth of service jobs is part of a world-wide trend.
- The rapid growth of the industrial economies of East Asia, together with relocation there of EU TNCs, has in general been exaggerated as a threat; however, the growth of imports from poor countries may account for the deterioration in the income and job prospects of the EU's unskilled workers.

3

THE CHANGING MAP OF
THE EU ECONOMY

In the light of Chapter 2, I shall now concentrate on the map of levels of prosperity within the EU. This will be by reference to data on employment, unemployment and the value of production in constituent regions, and incorporating three more regional case studies of contrasting areas. The effect of external trade on specialised, individual regions can be seen in many of the traditional problem areas of the EU. However, the rise of social exclusion in cities and the insecurity even of middle-class jobs can be seen in high unemployment in metropolitan areas. A study of the dynamics of change in two of the EU's most prosperous areas leads on to studies of change across the EU area as a whole, raising the question of whether regional differences are increasing or decreasing, before concluding with a special study of the most critical new problem area, Eastern Germany.

THE MAP OF EU UNEMPLOYMENT

Traditional industries

The last chapter concluded on the *ongoing* threat of world competition to Europe's more vulnerable workers:

> To say, however, that the problem is one of adjustment is to miss the details already etched on the map of Europe where lower-paid and women workers in industries vulnerable to international competition already compose a large part of the map of unemployment in Europe.
> (*The Economist, Supplement*, 1 October 1995: 29)

It is worth pausing to stress the links between the external relations of the EU, considered in the last chapter, and the internal map of human problems of unemployment, which occupies much of this chapter. It is possible, indeed, to relate global trading trends with the human situation in individual EU regions. Thus CEC (1994a: 50) point to the loss of 900,000 jobs in textiles and clothing in the early 1980s, affecting many weaker regions of the EU, 'largely as a result of the relocation of production to lowcost countries as well as the introduction of new technologies'. Global recession in metal industries was still reflected in 1995 unemployment in respect of a large area of weakness in the ex-Franco-Belgian Coalfield (with rates of 15.3 per cent on the French, and 15.9 on the Belgian sides), in other steel and coal areas such as Strathclyde (9.7), South Yorkshire (11.2), the Saarland (9.1) and the Ruhr (8.2), as well as in some shipbuilding areas such as Bremen (10.6) and Liguria (Genoa, 10.9).

The regional map

Figure 3.1 provides a snapshot in time of unemployment rates across the whole of the enlarged EU area, as standardised by CEC, Eurostat (1995a). The relative values for different areas tend to change only very gradually (though that is a point for a later section), so that at a different point of the cycle the same map might just as well apply, but with different levels of unemployment. A glimpse at the map will suggest that national boundaries play a role, whether because national economies fail to synchronise with each other over the economic cycle, or because differences of definition of the unemployed have proved incapable of adjustment by Eurostat.

The scale of regional variation, long recognised by the European Commission, also varies itself *between countries*. As shown by both the map in Figure 3.1 and the bar graphs in Figure 3.2, the depth of unemployment is far greater in Spain and Southern Italy than in the rest of Europe. The rate of unemployment exceeded 20 per cent in regions of Southern Italy, including, as we saw earlier, Campania at 23.1 per cent; yet Spain as a whole recorded a higher rate, one of 24.4 per cent, with a peak in the southern province of Andalucia with 34.7 per cent. These variations correlate with those of Gross Domestic Product per capita, but also have to be seen alongside the prevalent patterns of informal work in the area (Stratigaki and Vaiou, 1994) and the most important tendency for a growing population, and women seeking work for the first time, to push up unemployment.

Not only the broad lines of Southern Italy or North-East England tend to persist from cycle to cycle, but many more details of the map. This is partly because long-term 'structural unemployment' can result from the loss of agricultural, fishing, quarrying or industrial jobs. There is no doubt that the loss of 40 per cent of the UK's industrial jobs in 1971–91 is the main feature etched across the country's map of unemployment; individual counties' increase of unemployment in recession conditions was a direct function of their dependence on manufacturing jobs (Townsend, 1983).

However, the map has several striking new features. A later section will have to review the unique experience of the labour force of East Germany, whose former area is clearly etched out with high unemployment. The addition of East Germany to the EU in 1990, with its new labour market in turmoil, preceded the doubling of unemployment to unprecedented levels in

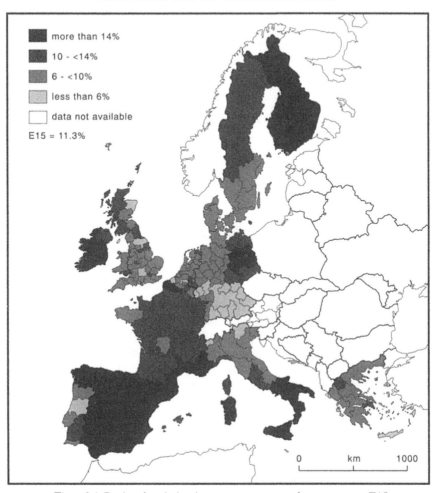

Figure 3.1 Regional variation in percentage unemployment rates, E15
(April 1995; both sexes)
Source: CEC, Eurostat, 1996a

the rest of Germany (1990–4). What was most surprising was the revolutionary doubling of rates in the same years before two new member states joined in 1995, in Finland, a change connected with instability in trade with the new Russia, and in Sweden. The welfare state of Sweden had prided itself on a high rate of economic growth coupled with good welfare and employment conditions. As recently as 1991 it recorded an unemployment rate of just 2.7 per cent, yet this had come to exceed the UK rate by 1994, reaching 9.2 per cent by early 1996.

The map of female unemployment (Figure 3.3) shows higher values than for males virtually everywhere except the UK, where financial gains to the

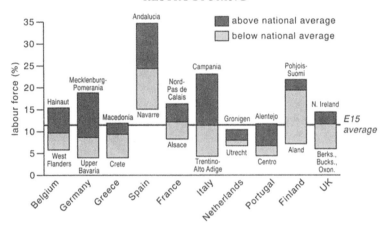

Figure 3.2 The range of regional variation in percentage unemployment rates by
member state (April 1994)
Source: CEC, 1994b

employers of part-time females and limitations on women's eligibility to
draw benefit are likely to explain the departure from the general pattern. The
recorded difference stands at its greatest in Spain, where on average nearly a
third of women are recorded as unemployed, and over 45 per cent in one
region! The male rates for Campania and the British unit of Cleveland and
Durham are similar, 18.4 and 14.8 per cent respectively, yet the contrast in
female rates could hardly be greater, between 32.3 and 7.5 per cent
respectively.

Metropolitan unemployment

Issues that we thought were confined to problem regions of the periphery
are now the hallmark for much of Europe, with 1995 rates of unemployment, for instance, of 12.0 per cent in Greater London, 13.3 in Brussels, 11.2
in Berlin or 10.0 in Paris (CEC, Eurostat, 1996a), meaning more than one in
ten of those in work or actively seeking it.

There is international evidence that the 'excluded' households we found in
Chapter 2 are spatially concentrated to an increasing and critical extent,
crucially in cities. Mingione (1993) has contended that the scale of segregation is increasing, with large European cities becoming Americanised and
divided into clearly separate areas of affluence and 'ghettoes' of chronic
poverty and marginalisation. This has led Musterd (1994) to question
whether we have 'a rising European underclass' created under social polarisation and spatial segregation. In the USA, it has been argued that if such areas
of extreme poverty are spatially concentrated rather than dispersed, then the
disadvantaged within them will become isolated from the economic and

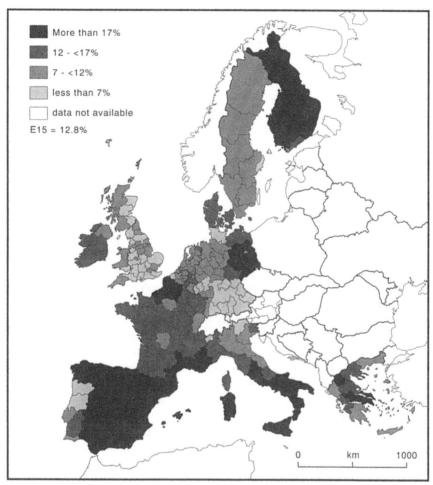

Figure 3.3 Regional variation in female percentage unemployment rates, E15
(April 1995)
Source: CEC, Eurostat, 1996a

social mainstream. That is, a concentrated group will have less contact with other groups than if all were evenly distributed; evidence from the long-term unemployed (Griffin *et al.*, 1992; Dawes, 1993; Morris, 1993) reveals that their social networks consist disproportionately of other unemployed people, so isolating them and potentially reducing the number of employment opportunities available to them.

A further factor affects the labour market of cities. Swedish restructuring has demonstrated for the first time in the industrialised world a restructuring due to major job loss in services, which was accompanied by reactions in other parts of the labour market and economy. Elsewhere, the effects of the early 1990s recession resulted, for example, in a downturn of numbers of workers commuting into central London, reversing a late 1980s increase. Corporations such as ICI and BP have reduced the volumes of their central London offices, while in Paris, Société Generale, the bank, is leaving many smaller, older buildings to occupy a new headquarters block with less space in la Defence, the new business district (*Financial Times*, 3 April 1995: 8). (The more stringent management of office workspace is accompanied by 'hot-desking', under which sales and other mobile workers have no permanent desk, but are assigned one for the day from a limited stock when they do attend; an example in the public sector exists in proposals of August 1996 by Birmingham Corporation.)

REGIONAL VARIATIONS IN PRODUCTION

For demographic and cultural reasons which we have given, disparities in unemployment somewhat exceed those in regional production. The CEC (1991a) argue that unemployment disparities are about three times higher than in the (longer-established single market of the) USA, whereas those in regional GDP per capita are only twice as high as in the US. At the national level, Greece and Portugal had levels of only around 63 and 69 per cent respectively of the EU average in 1993, when expressed in terms of parity of purchasing power (rather than taking currency exchange rates at face value; CEC, Eurostat, 1996b). On this basis, most countries cluster at between 102 and 113 per cent of the European average.

Figure 3.4, however, shows more clearly than the unemployment data, the existence of an international core of more prosperous areas extending from Northern Italy to Paris, West Germany, Belgium and South-East England. For example, there was a small group of regions with GDP per capita of more than 25 per cent above the EU average, including London (143), Brussels (174), Baden-Wuerttemberg (131), Bavaria (126), Eastern Austria (128), Emilia-Romagna (128) and Lombardy (134). Like the areas of lowest national unemployment at Figure 3.1, these tend at their southern end to provide a ring around the Alps, while the EU's poorest areas are dispersed around the periphery, in Portugal, Greece and now East Germany.

This core-periphery pattern is very well recognised, and is linked by Amin and Tomaney (1995) to European unification, through the concentration of investment and scientific industries near the commercial centre of gravity of the unified single market. At any one time, there are bound to be regions or sub-regions that are gaining or losing relative to the EU average. The questions for this chapter are whether these changes have represented basically the sectoral and occupational trends that are the theme of the book, whether there is a cultural or institutional impetus to change, or whether there is an overall spatial pattern arising from the imposition of a Single European

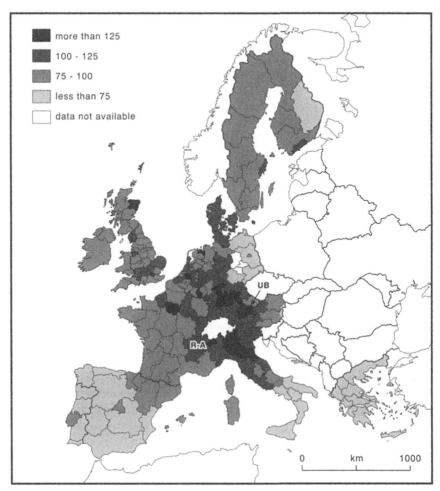

Figure 3.4 Regional variation in the value of production at common purchasing power
standards, E15, 1993, E15=100
Note: R-A = Rhône-Alpes
UB = Upper Bavaria
Source: CEC, Eurostat, 1995c

Market. In fighting competition with other continental trading blocs, such as
the United States of America, economic integration may be widening disparities within the territory of the EU.

I have therefore chosen to study two of the regions surrounding the Alps,
Rhône-Alpes ('R-A' in Figure 3.4) and Upper Bavaria ('UB') to assess what is
the nature of their recent dynamic, before meeting the new and important
need to consider the special case of East Germany.

57

TWO SUCCESSFUL CORE REGIONS

Rhône-Alpes

The concept of a 'Europe of the Regions' implies some independent dynamic for provincial areas *vis-à-vis* their respective national metropolitan regions, even in that most traditionally centralised country of France. Here the very lowest rates of unemployment are found adjoining the German border in Alsace-Lorraine, and much of the modern growth of business may still be associated with the traditional economic core of western Europe defined by Paris, London, Amsterdam, the Ruhr and Frankfurt. With a shift in prosperity to EU regions bordering the Alps, an outstanding area of growth and prosperity is found in Lyon and the large belt of territory lying between the River Rhône and the Swiss border, known as Rhône-Alpes. With 5.3 million people, it had longstanding experience of industry in pockets such as Lyon and St Etienne, but industry itself has given way to a strong service dynamic since the late 1960s.

The human experience of change in this area is one of expanding population, housing, training and work, increasingly associated with attractive rural environments and improved infrastructure. The area stands well ahead among provincial regions through investment in advanced industry supported by advanced business services. Over 12,000 workers are employed either in the Renault vehicle chain or in Rhône-Poulenc chemicals, supported by a wide range of small and medium scale industries. Two strong features that add to the area's high salaries are the growth of research and of business services. R&D laboratories in both the public and private sectors are particularly connected with the University of Grenoble and the usually credited reputation of an attractive living environment for graduate staff. Business services appear to have expanded beyond the critical size at which specialist business partners are available for the flexible development of new business (Mayere and Vinot, 1991), sufficient for 43 per cent of all sales of this sector to be made beyond the region: 15 per cent to Paris and 7 per cent to other countries. Service growth has been slowed little by recessions and the sector now employs more than 60 per cent of all workers of the area. With the assistance of communications improvements, motorways, airports and the TGV (Train à Grande Vitesse) to Paris and Marseille, with a projected extension to Turin in Italy, Lyon is selling itself in competition with surrounding cities, notably against Geneva in Switzerland. However, it appears to be sharing in a widespread regional growth of business services, which are, for example, decentralising widely from Geneva to accessible areas of French-speaking Switzerland (Boulianne and Decoutere, 1994).

The case of Bavaria, in southern Germany

The area that stands out on the EU map for its shortage of labour (Figure 3.1) is undoubtedly southern Germany, with Bavaria as the EU's region of lowest unemployment, and Upper Bavaria, which includes the city of Munich, as the lowest of any area quoted by CEC, Eurostat (1996a), with a rate of 4.1 per cent in 1995.

It was not always so, as the area was still markedly rural in the 1950s. Population growth became concentrated in Bavaria, and in Baden-Württemberg, where the role of cultural, social and institutional factors in economic development has been emphasised by Cooke and Morgan (1994). In Bavaria, the symptoms of the Nord–Sud Gefaelle (north–south divide) lie in the explosive growth of the Munich agglomeration across the map and the experience of lengthening journeys to work to dynamic employment centres. An economic upturn from 1984 led to a steady drop in unemployment to a low point of 3.4 per cent in 1990, bringing levels below 2.0 per cent in parts of the Munich conurbation and attracting foreign migrants who now form 20 per cent of the population. The labour shortage also prompted long-distance commuting, so that by the late 1980s more than a third of the 800,000 working population of Munich were commuters (Helbrecht and Pohl, 1993).

How do we explain the industrialisation of the area and the rise of Upper Bavaria's income per head to 41 per cent above the E12 average, almost the highest in Germany? One contributory factor lay in regional assistance for peripheral rural areas near the Iron Curtain, leading to expanding education and a structural revival in frontier areas. A more fundamental reason appears to lie in the very industrial virginity of the area:

> During the post-war period, because of the lack of primary and heavy industry, the Bavarian economy was able to adapt more readily than the old industrial areas to the new high-growth industries such as the capital goods industry.
>
> (CEC, 1993a: 20)

In the EU as a whole a number of non-coalfield cities established themselves in the electrical engineering sector, and Munich is notable for having over a quarter of a million jobs in this sector, including post-war growth by the leading Siemens group; mechanical engineering and vehicles each have about 200,000 workers, including BMW and Audi. These three sectors account for 56 per cent of international exports from Bavaria, more than half of these going to the rest of the EU. Following rapid technological growth and high levels of technical training, a shift to services has been notable, including the growth of a few financial companies, such as Allianz, which operate on an inter-regional and international basis. Perhaps reflecting the longstanding and continuing demand for staff in industrial firms, however, the proportion of employment in services, even now in Upper Bavaria, is below the levels obtaining, for instance, in France. Thus we see that national features, and probably cultural factors, will complicate any firm explanation of the location of the highly successful economies of the EU.

CONVERGENCE OR DIVERGENCE IN ECONOMIC TRENDS?

The traditional problem areas of the European Community in the past lay at the greatest distance from Rhône-Alpes and Bavaria, in the 'periphery' rather than the core:

> In the 1970s, the European Community's 'regional' element was a mission to the 'non-affluent', intended to keep them harmless and, if possible, at home. Even when seen positively, the 'regional issue' tended to formulate itself as the 'historic' regions – Scotland, Brittany, Corsica, etc. – versus their respective central states.
>
> (Harvie, 1994: 52)

This quotation reminds us how the Community inherited from member states a set of problem regions whose disadvantages certainly included those of poor accessibility and remoteness from markets. However, it is generally agreed that, with some help from European funds, there was a period of 'convergence' up to the mid-1970s, meaning that the gap in economic activity narrowed between the wealthiest and poorest regions. CEC subsidies helped national ones in bringing new industry to, say, Scotland or Southern Italy, reducing the economic differences between those areas and the rest of the Community's area.

The 1980s saw the arrival of a new kind of 'regionalism', in which areas such as Bavaria, Baden-Württemberg, Rhône-Alpes or Catalonia sold the concept of the 'Europe of the Regions with all the energy one might expect of such well-heeled areas' (Harvie, 1994: 62). Given the clue that defensive self-interest was at work in the political discourse of both the core and the periphery, we may go on to ask who has been winning; have the economic and policy forces at work promoted convergence or divergence?

Changes in regional GDP

Although I am writing primarily about people's experiences of employment, I shall first review changes in overall output both at the national and regional levels. A slight divergence of member *states'* levels of output from 1980 to 1984 was followed by an undoubted period of convergence at a time of widespread higher growth:

> During the late 1980s overall rates of economic growth in the four

least developed member states were higher than the EC average, and this translated into an improvement in GDP per head in Spain, Portugal and Ireland, although this was not the case in Greece.

(Tomaney, 1994: 161)

It may also be noted that the newest, 1995, members of the EU – Sweden, Finland and Austria – meanwhile lost their average 10 per cent advantage in production entirely in the period 1985 to 1992, converging on the EU average GDP per capita before joining.

Disparities in country levels of GDP per head may be measured in terms of the standard deviation of the EU average (after weighting for size of population). Figure 3.5 clearly shows that the period from 1984 to 1993 ushered in reduced disparities at this level; 'contrary to some worries, the peripheral member states appear to have benefitted more than proportionately from integration' (Reichenbach *et al.*, 1994: 141), and the timing would allow some credit to the tariff cuts resulting from the Single Market.

However, this line of the graph is not matched by the one above for disparities in GDP per head *by region;* indeed this data from the CEC itself can be read as showing a worsening from 1980 to 1991; at any rate the poorest twenty-five regions showed no change in their average output per head from 1980 to 1991, 2.5 times below the EU total. But, as stated by CEC (1994b: 41), 'the real challenge, however, is one of ensuring that productivity gains are accompanied by output growth allowing employment to increase and unemployment to decrease'.

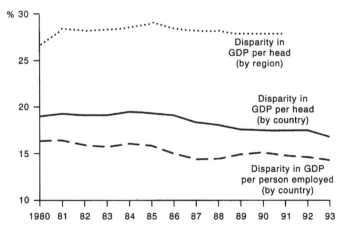

Figure 3.5 Regional and national disparities in the per capita value of production, E12, 1980–93

Note: Disparities are measured as the standard deviation, among units studied and weighted by population, as percentage of the average EU GDP per capita

Source: CEC, 1994b

Change in employment and unemployment

'It is significant that regional differentials in productivity, unemployment and employment appear to have *increased* in the last decade, the period in which efforts to integrate the European economy have been most intense' (Amin and Tomaney, 1995: 10, emphasis added). There are several reasons why the picture is worse when we turn from production to employment, as demonstrated for 1994 in Figure 3.2. Given high priority for their lagging economies, attributed particularly to their poor infrastructure and training levels, many of the industrial regions that have shown improvement in terms of GDP have achieved this primarily through increases in productivity (CEC, 1994b); thus the effect on employment has not been commensurate. When we view employment change in Southern Europe, we see that Spain was achieving net job gains little faster than England and Wales, when there were marked increases in the demand for work.

Relative disparities in unemployment have therefore risen during the large absolute increase of joblessness from the 1970s to the 1990s. Figure 3.6 shows that the temporary improvement in unemployment achieved across the EU from 1984 to 1990 failed to check more than temporarily the rise in regional disparities, which widened again in 1992 and 1993; in 1993 the ten worst regions averaged an unemployment rate of 25.3 per cent, and the ten best one of 3.6 per cent.

Divergence, a picture of the worst getting worse and the best relatively better, is shown in Figures 3.7 and 3.8 covering the period 1987 to 1994. It can be argued that our central belt of prosperity extending from South-East England across most areas of Benelux, northern and eastern France to

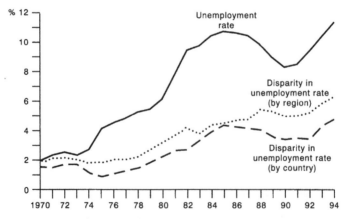

Figure 3.6 Regional and national disparities in unemployment rates, E12, 1970–94
Note: Disparities are measured as the standard deviation, among units studied and weighted by population, as percentage of the EU average
Source: CEC, 1994b

northern Italy fared relatively well from 1987 to 1990; 'there is important evidence that the historical dominance of this core is being reinforced by changes in the nature of production which place a greater premium on scientific and technical innovation' (Tomaney, 1994: 65).

The maps reflect, pending the accession of East Germany, Sweden and Finland, that the principal issue lay in a worsening of problems in Mediterranean Europe, including not only a disappointing worsening in Greece and Southern Italy in the first period, but also a later deterioration in parts of Southern France and across Spain (as well as South-East England and northern France). 'Unemployment in the weaker regions fell by less than the number of new jobs, and by less than in the stronger regions' (Collier, 1994: 148). The reason lies in demography, in the greater numbers of young people attempting to enter work in these areas, and also in the basic cultural point about female activity rates rising from a low level.

Peripheral areas of the EU appear to have specific difficulty in facing international commercial competition. In general, small and medium size enterprises

> tend not to possess the resources required to respond to rapid techno-
> logical change in increasingly global markets and the development of
> new products of ever higher quality which it involves. The problems
> are more serious for firms in the weaker regions.
>
> (CEC, 1994b: 38)

In addition, surprisingly few areas have done as well as the UK in attracting investment from outside the Community (e.g. from Japan or the USA). Mergers of European companies to meet the demands of the Single Market may further concentrate their best paid jobs in the central regions of the Union.

In concluding the whole volume, Chapter 11 will include reference to the EU's new Structural Policies of 1994, incorporating specific new problems in the shape of a totally new kind of problem area, that of East Germany. At this stage, however, in relation to policies for the old E12, we may conclude, with Collier (1994: 146) that 'there is no evidence that regional convergence is happening in Europe. Such convergence as took place in the post-war period only did happen because of the differential growth rates of rich and poor countries.'

The example of East Germany

East Germany is a most important addition to the EU with regards to this chapter – both in its own right and as a possible precursor to the accession of other 'transitional' economies from Eastern Europe in the future. On a superficial glance at our maps in the last section, East Germany fits a pattern of low output per head and high levels of unemployment shared with other

Improvement in unemployment rates 1987-90

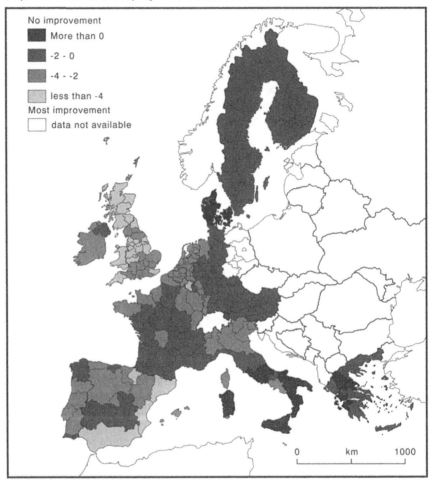

Figure 3.7 Regional change in unemployment rates, E15, 1987–90 (improvement)
(percentage points)
Note: No regional data available for Sweden, Finland and Austria
Source: Commission for the European Communities, 1995b

peripheral regions of Europe, which we might attribute to that very periph-
erality. However, the reasons for its severe economic difficulties lie in its
transfer from one continental and political trading bloc to another in 1990; in
fact from being, as the German Democratic Republic (GDR), the wealthiest
country in per capita terms of the Communist Soviet bloc, to becoming by
far the poorest region of the EU's wealthiest state, and at one stage, in 1993
data (CEC, Eurostat, 1996b), showing lower production per capita levels
than all other EU regions. The EU therefore has acquired a problem of

Deterioration in unemployment rates 1990 - 94

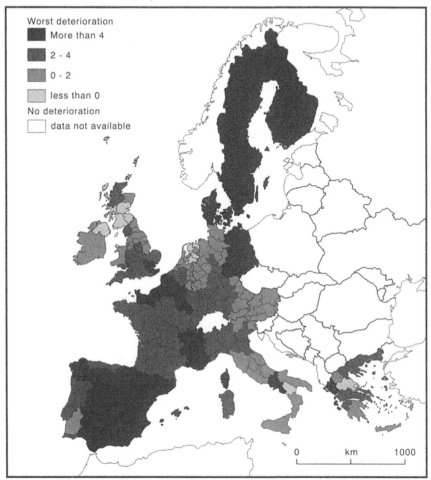

Figure 3.8 Regional change in unemployment rates, E15, 1990–4 (deterioration)
(percentage points)
Note: No regional data available for Sweden, Finland and Austria
Source: CEC, 1995b

about the size of Southern Italy, while the new Germany (Jones, 1994) has to heal a divide as great as persists between the Mezzogiorno and the rest of Italy.

Despite a continuing loss of population to the West, GDR central planning had been successful enough in building on longstanding geographical zones of specialised industry for the country to challenge the UK in levels of output by the early 1980s, before a diminishing rate of growth made the country more open to winds of change blowing throughout Eastern Europe, which lead to the fall of the Berlin Wall in 1989, and voluntary accession to the existing German state in October 1990.

A visit to the area five years later showed much of it to be a building site, so great is the scale of reconstruction. Rutted roads, cobwebbed closed factories and dusty, time-worn town centres and government-built flats are giving way to enhancement of streets and housing blocks, but above all to a private sector feast of new out-of-town developments comprising garden centres, do-it-yourself centres, McDonald's outlets, petrol stations, housing and new golf courses.

People's lives have been through an even greater series of changes, not all for the better. Industrial, water and soil pollution are now improving, partly through closures in heavy industry. On the other hand, employment in East Germany dropped by nearly 40 per cent between the first half of 1989 and the second half of 1993, a reduction smoothed only by temporary employment measures including short-time working and public work programmes. The body that was responsible for the privatisation of all East German enterprises, the Treuhandanstalt, also cushioned the blow by injecting capital into certain businesses in order to avoid closure, thus keeping employees on its books albeit temporarily.

Otherwise, from the day of financial integration with the West, and the loss of many trading arrangements with former Eastern bloc countries, establishments in East Germany were left to sink or swim. The general view was that, due to outdated technology, methods and marketing, only about a third of East German production could continue in full exposure to EU market forces, another third would need temporary subsidy, and a third would have to close. In the event, the East German consumer showed an immediate and strong preference for Western goods, and GDP per head fell from two-thirds of the EU average in 1988 to a third in 1991, before recovering to a level of a half by 1993 (CEC, 1994b). From 1989 to 1993, over 2 million of the 3.6 million East German manufacturing jobs were lost through redundancy, together with some 70 per cent of farm jobs (Jones, 1994).

As stressed by the CEC (1991a: 238), East German activity was concentrated in the 'material production sectors', in energy and transport as well as manufacturing and farming, but with over half the industrial output in mechanical engineering, electronics and chemicals. Redundancy and investment produced a sharply British pattern of deindustrialisation, with investment largely concentrated in services. Agriculture, mining and energy and manufacturing each lost employment, typically about a half, in all six constituent administrative areas (Laender) of East Germany; the only sector to gain jobs in all areas was that of banking and insurance (which doubled from a small size), although construction showed very significant gains in most areas.

We would not normally expect such a sectoral change to be adverse to women, but in fact their share of jobs fell from a remarkable 49 to 43 per cent, amid general agreement that women have been the most affected by the contraction of the labour market in the new Laender (Jones, 1994; Quack and Maier, 1994; Rudolf *et al.*, 1994). Quack and Maier argue that women suffered proportionately more from job loss because of the sectoral distribution of closures, and because of the previous division of labour within establishments in general: women were over-represented in those industries that closed down most rapidly, primarily textiles, electronics and food processing, while trade union intervention slowed the progress of closures in male-dominated sectors such as steel and shipbuilding; women were over-represented in those occupations that proved to be over-staffed by West German standards, for instance in the administration, clerical and social work of factory complexes.

Thus, despite migration, commuting to the West, and the disappearance of women and older workers from the labour market, women's recorded unemployment grew more rapidly than men's. In Figures 3.1 and 3.3 (above), the ex-GDR stands out for some of the highest unemployment rates of northern Europe, with 1994 female rates in excess of 20 per cent everywhere except Berlin, and male rates greater than 10 per cent in the majority of areas. Viewed in more detail, the larger agglomerations tended to show the lower rates, as in Leipzig.

Over three-quarters of state financial transfers from the rest of Germany are spent on the personal recurrent costs of providing unemployment benefit, pension, health care and other welfare benefits. This has only delayed the permanent investments intended for urban improvements, environmental correction measures and infrastructure, as well as re-training itself, all under a hierarchy of regional plans. There has been a growing realisation that the task is more far-reaching and difficult than felt in the heady days of 1989. The task therefore of restoring normal living conditions in East Germany began to tax the patience as well as the pockets of West German voters. There are indeed fears that East Germany may remain like Southern Italy, as a permanent drain on public and EU funds. Clearly in regional planning terms, there are few precedents for closing a gap as big as that between the two parts of Germany. A professional judgement must be that this would take the better part of fifty years. However, it could be that a greater migration of industry to the East of the country would help to contain the rise of German labour costs, as the main threat to the country's industrial survival.

SUMMARY

- The European map of regional unemployment has long shown the effects of industries growing on a global basis, notably textiles and steel.
- The overall map of unemployment in the EU still tends to display a peripheral pattern of high unemployment in Greece, Ireland, etc., but this is now joined by Sweden and Finland.
- There is a central belt of greater prosperity and lower unemployment,

from South-East England to northern Italy, within which regions such as Rhône-Alpes and Bavaria appear to gain from cultural factors and their relative lack of the social and environmental problems of older industrial areas.

- There has been 'convergence', an evening out of disparities between member states' economies, in the 1980s, but this has not been translated into greater regional equality in unemployment, where levels have tended to diverge.
- The accession of East Germany has added a problem as great as that of Southern Italy which is likely to take several decades to alleviate.

Part II

THE MAIN SECTORS OF CHANGE

In this section the major changes in the sectoral and gender restructuring of employment are examined in turn. As the book is concerned with *change* in employment in Europe, industry is regarded as a negative element, because the prevailing trend is quite clearly for industrial jobs to decline in number. In framing the approach to the book at the end of Chapter 1, Figure 1.5 showed that a minority of industrial sectors did expand their jobs across the EU from 1985 to 1990, but hardly any from 1990 to 1994. Thus, the actual expansion of direct industrial employment (as opposed to that of production) is now limited very much to particular industries in particular regions of countries at particular points in the economic cycle; the main exception in the year of writing (1996) is the attraction, led by the UK, of FDI from Japan, Korea and Taiwan to greenfield sites intended to employ 2,000 to 6,000 workers.

A declining industry in terms of employment can still generate job growth in relevant services and suppliers from the general increase in production, but even those may come from other regions or sub-regions.

Conversely, the majority of attention of this section is given to the expansion of service activities and women's employment. The first chapter, concerned with 'deindustrialisation', is followed by five on expanding sectors (Figure 1.5) and their geographies.

4

FACTORY JOB LOSS – GLOBAL SHIFT OR DEINDUSTRIALISATION?

> The kind of world that I'm studying – what corporations do and why they do it – is populated by strategic actors . . . I think you need to talk with them and find out the story.
>
> (Schoenberger, quoted in Hodges, 1995: 17)

In this chapter I shall begin by looking at the individual industrial corporation and its role in taking away jobs from industrialised regions, factory job expansion in the EU being fairly rare outside small new plant, as explained on page 69.

Although we established in Chapter 2 that the role of TNCs in transferring existing or new investments to other countries had been exaggerated, at least at the macro-economic level of the EU, it does not follow that it may not be significant at the level of the individual town where people live, or the smaller regions whose unemployment was mapped in the last chapter. The principal question to be established in this chapter is the significance of corporations' international restructuring, or whether job loss in the EU is of a much more general kind, to be defined and statistically analysed under the general heading of 'deindustrialisation'.

WHY STUDY CORPORATIONS?

As in the last chapter, local outcomes and characteristics cannot be read off simply from aggregate structural trends. The effect of new directions of economic change on any one area is mediated through a range of contingent circumstances, usually including the role of the state and of large corporations, whether in the public or private sector. In particular, case studies have shown that the 'strategic actors' in different corporations have differing degrees of choice in meeting problems, for instance those of overcapacity. Their potential power is immense; as has often been remarked, individual corporations have bigger transactions than many medium or small countries. For example, transactions would rank the US vehicles firm General Motors twenty-third among countries of the world, ahead of Austria, Sweden and Switzerland, and of all except four low income countries.

For any individual locality, this play of choice may make all the difference between the expansion or full closure of the local branch of a multi-plant firm or TNC (Fothergill and Guy, 1990). This decision is, however, influenced by the organisational framework of the firm (Peck and Townsend, 1984, 1987). Let us assume that a local planner or politician (and I am both) wishes to understand the likely future dynamics of the local branch of a TNC. He or she needs not just to identify the statistical category in which the plant falls, but to study the economic pressures on the financial decision-making unit of which it is a part. Which division of the parent company controls the plant? How close is it to the central head office? When was it acquired? (More recently acquired subsidiaries and branches have been found more vulnerable to closure in both Britain and the USA; Howland, 1988).

THE SIZE OF WORLD COMPANIES

TNCs are responsible for a fifth to a quarter of world production and are the single most important agent in creating global shifts in economic life (Dicken, 1992). I am following Dicken in defining a TNC as a company with assets in more than one country, not necessarily spread across a greater range of countries in a 'multinational' way, as only 4 or 5 per cent of TNCs are truly global in scope. Nonetheless we saw in the last chapter that the total direct employment of TNCs in the mid-1980s was put at 73 million.

The basic advantages of the large firm are regarded as those of ownership; the first in the shape of exclusive or preferential access to a particular market; the second in the shape of a unique asset such as a patent or trade mark; and the third in the ability of hierarchies to organise related activities more efficiently than the market.

European TNCs

New data for Europe as a whole (*Financial Times, FT500* supplement, January 1995) shows the greatest concentration of assets in international producers of crude oil, with an average value per company of $18,539 million, followed by recently privatised telephone companies, the motor vehicle industry, chemicals and drugs, natural gas and electrical equipment, in that order.

These are represented among the leading twelve European firms by employment shown in Table 4.1. This ranking relates to three broad types of manufacturing in TNCs: (1) technologically more advanced sectors, (2) large volume, medium technology consumer goods industries and (3) mass production consumer goods industries supplying branded products. In Table 4.1 we see the largest employers of Europe dominated still by manufacturing firms (particularly from Germany), by vehicles, electrical engineering and by makers of branded food goods, but the leading twenty-five demonstrate the replacement of many manufacturing firms by service activities. Two leading UK firms, ICI and GEC, have fallen out of the top twenty-five, along with the big three German chemical firms, Bayer, BASF and Hoechst, while 'holding companies' with a miscellany of assets entered the list. Despite the effects of the Single European Market and of a wave of European mergers in the mid- and late 1980s (Figure 4.1), there are still relatively few genuinely

Table 4.1 Europe's leading employers, including world-wide operations, 1994

Company	Country	Sector	Employees
1 Siemens	Germany	Electrical engineering	391,000
2 Daimler-Benz	Germany	Vehicles	371,100
3 Unilever	Netherlands/UK	Food and oils	294,000
4 Fiat	Italy	Vehicles	261,000
5 Volkswagen	Germany	Vehicles	253,100
6 Philips	Netherlands	Electrical engineering	252,200
7 Nestlé	Switzerland	Food processing	209,800
8 ABB Asea Brown Boveri	Sweden/Switzerland	Electrical engineering	206,500
9 Générale des Eaux	France	Water	204,300
10 Alcatel Alsthom	France	Communications equipment	196,500
11 BAT Industries	UK	Tobacco, food, etc.	190,300
12 Hoechst	Germany	Chemicals	172,500

Source: Financial Times FT500, January 1995

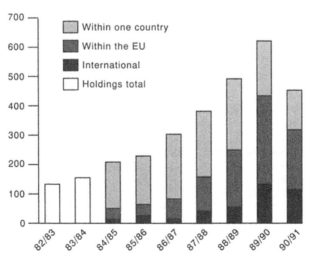

Figure 4.1 Mergers and acquisitions in the EU, 1982–91
Source: Global Labour 1 May 1993

73

international companies like the long-established Anglo-Dutch enterprises of Unilever and Shell.

EU corporations' relative investment in one middle income country: Brazil

To compare Europe's external investment, I was much interested to examine the stock of externally owned TNCs in one large middle income country, Brazil. The practice of ranking 500 leading companies has spread from the northern hemisphere to provide similar data (*Conjuntura Economica*, 1994; although in this case no employment figures are available). Table 4.2 shows from a general ranking how the total activities in Brazil of externally owned corporations stand among domestic corporations. The second largest corporation of Brazil is the US Ford Motor Company's manufacturing and distribution subsidiary in the country, and three more US corporations appear in the top twelve foreign entities.

However, there are eight European-owned operations from all the leading European countries in the top fifty. In Brazil, seen in the 1970s as the 'darling of the international business community' (Chomsky, 1994: 221), much of this development occurred in large Fordist branch factories, applying the corporation's standard routine methods to large-scale cheap production, only in this case for the domestic markets, attracted by high tariffs. At the time little discernible effect would be seen on home plants in the industrialised world; it was a question of differential growth, *but* giving way to more widespread 'global shift'. Absolute job losses followed in the home country, though not usually of white collar staff, as when Volkswagen, for instance, moved its export production of cars from Germany to Brazil.

Table 4.2 Transnational corporations in Brazil, ranking among all corporations (excluding public sector), 1993

Ranking (general)	Company	Parent (named if different)	Activity
2	Autolatina Brasileiro	Ford Motor, USA	Vehicles
6	Souza Cruz	Imperial Group, UK	Tobacco, misc.
7	Fiat Automoveis	Fiat, Italy	Vehicles
10	Mercedes Benz	Germany	Vehicles
16	Rhodia	Rhône Poulenc, France	Chemicals
20	Shell	UK/Netherlands	Oil
22	White Martins Gases	USA	Chemicals
39	Dow Quimica	Dow Chemicals, USA	Chemicals
40	CIBA-Geigy	Switzerland	Chemicals
41	Makro Atacadista	Switzerland	Oil
44	Texaco	USA	Oil
48	Mannesmann	Germany	Engineering

Source: *Conjuntura Economica*, Rio de Janeiro 48, 8: 48–50

The relative position of the European base and poor country branches is now altogether more fluid. There are cases of complete transfer reported, when these in reality were rare. In a complete scanning of the *Financial Times* from 1976 to 1981 for Townsend (1983), I came across only one full transfer, when in fact a Brazilian firm, having acquired the Swindon firm Garrard, which made record changers, announced in 1982 that it was transferring the remaining production to the Manaus Free Trade Zone on the Amazon.

Recent change in FDI to low income countries

More recently, British Polythene Industries, Europe's largest polythene film producers, decided to close its factory in Telford New Town, Shropshire, losing 150 jobs, in order to enter a joint venture with a Chinese state-owned firm in Guangdung province; 'our entry into China will now give us a chance to compete on our competitors' terms' (*Financial Times*, 7 February 1995: 7). In the USA, the Resource Center Bulletin (1993) focused on 'runaways', where manufacturing sites are emptied after transfer of production abroad, above all for the purpose of slashing the wage bill. They document the relocation of more than 96,000 jobs from more than 250 work sites in the US over twelve years. The threat of such movement is used by firms to bring their workforce and unions (if any) into line over pay and conditions.

However, evidence is now increasing that TNC jobs in poorer countries are no longer all low paid. Undertaking research for the European Commission, Howells and Wood (1993) describe how companies have elevated the role of units around the world through a more decentralised and open management system, through 'global networking' and 'global switching'. As from the 1980s, TNCs began to transfer R&D activities on an international basis, and in this connection we see the growth of high paid jobs for men, and better conditions for women.

Effects within richer countries

Taking one country, the UK, as an example, it is possible to contrast the build-up to 67,000 jobs in 206 Japanese factories by 1994 with the downsizing of UK-based corporations, especially at their domestic plants. For example, ICI (Imperial Chemical Industries) reduced its workforce in the UK from 82,500 in 1976 to 47,300 by 1990 (Townsend, 1993), a reduction of 42.7 per cent, while its jobs overseas were increasing from 60,000 in 1976 (Clarke, 1985) to 78,400 in 1990. This major reversal was attended by automation in many divisions and by a greater rundown of manual jobs in the North than of white collar and women's jobs in the South (Townsend, 1993).

However, while some of the ICI investment was in poorer countries in the Commonwealth, for example, much was directed to other rich countries such as The Netherlands. The rich world's share of FDI at large actually increased to receive three-quarters of the world total in the late 1980s, with much more equal flows to and from the USA. Much of this investment is in the form of buying other firms, or 'acquisition activity', which fell away in recession conditions around 1991. However, improving economic prospects at the beginning of 1995 in Europe were enticing UK companies into a record level of spending on acquisitions (Figure 4.2).

All this cross-investment increased the size and proportion of TNC *foreign* employment in richer countries. *This was estimated in the mid-1980s at 22 million, compared with 43 million in home countries and 7 million in poorer countries* (UNCTAD, 1994).

TNCS' THREAT TO JOBS IN RICHER COUNTRIES

There is debate about the merits of 'foreign takeovers' in any country. President Clinton's Secretary of State for Labor, Reich (1991), develops the view that what counts is the location of work. Now that Americans are increasingly employed in America by foreign firms, this is wholly to the good, he argues, in providing jobs and introducing local workers to new technology.

This view will not necessarily hold in Europe, however. The pattern of foreign manufacturing investment in the UK does implicate external owner-ship in some of the adverse trends of the last two recessions, as trade unions have complained. The effect of this has been to reduce jobs in foreign-owned factories from 858,100 in 1981 to 774,800 in 1991, a drop of 9.8 per cent. This compares, however, with one of 23.5 per cent in the UK-owned private manufacturing sector. There are, of course, criticisms of other kinds about, for instance, the 'Japanisation' of British industry. One of the latest studies of new factories in Wales argues that Japanese manufactur-ers have contributed significant employment growth, and a diversification of products and capital assets (Munday *et al.*, 1995).

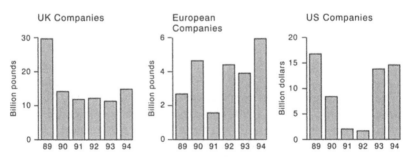

Figure 4.2 Mergers and acquisitions by UK companies, 1989–94
Source: *Acquisitions Monthly*, cited by *Financial Times* 4 January 1995

Recent downsizing by large corporations

The key point regarding TNCs and jobs is that they employ fewer people in richer countries as a whole than twenty years ago. Taking the United States' leading 500 companies as an example, their global employment fell from 16.5 million in 1980 to 11.5 million in 1994 (Hale, 1995). A leading firm, whose employment peaked at 405,000 in 1985, reduced its global employment by 1994 to 256,000 after a large amount of restructuring. That firm was IBM (the International Business Machines Corporation), which from its origins in 1911 expanded increasingly in Europe through its world dominance of the manufacture and sales of large mainframe computers, on the whole following the regional policy requirements of European governments (Kelly and Keeble, 1990) and avoiding any redundancies through redeployment.

By the mid-1980s, however, its products, management organisation and sales were being overtaken by new technology, with the growth and networking of miniaturised personal computers and new kinds of software. From the mid-1980s the corporation began to redevelop factories and to break down the previous divisional organisation through allowing more local autonomy and joint ventures, which Kelly and Keeble (1990) accept as general evidence for a 'flexible regime'. In the world recession, however, the firm made losses from 1990 to 1994 and had 50,000 redundancies in 1993. By 1995, it was back in profit but still facing tough conditions (*Financial Times*, 12 January 1995), which had also been shown by job losses in other computer manufacturers including Apple, Unisys and ICL.

Although computing is a field of rapid change, other industries are showing similar schemes of world 'rationalisation'. For example, Ford Motor Company merged its European and North American businesses into one from the beginning of 1995, aiming to reduce duplication in developing a new world range. Major restructuring and job losses were also reported in the EU by Vauxhall, Volkswagen, Renault and Peugeot. In other industries, international job losses were reported by Siemens, Unilever and Philips (which appeared among Europe's largest employers in Table 4.1) and by Shell, British Petroleum, Kodak and Rank-Xerox.

In the face of such changes, international union organisations expressed alarm, not least over the ability of TNCs to play off against each other the interests of different European countries. Writing in a new journal, *Global Labour*, Ramsay (1993) argued that TNCs' extension of interest from Southern to Eastern Europe heightened union fears of 'social dumping', in which TNCs forced countries to compete for investment by undermining progressive laws and higher standards.

Global downsizing by UK corporations

Embracing two heavy recessions together, the period from the financial year 1979/80 to 1992/3 demonstrates the massive global downsizing of the UK's

Table 4.3 The downsizing of leading UK employers; world employment as reported
(thousands)

Rank	Company	1979/80	1992/3	Change (%)	Remarks
1	General Post Office	411	385	−6.3	Now BT and PO
2	National Coal Board	297	68	−77.1	Now private
3	British Rail	244	138	−43.5	Since split
4	British American Tobacco	185	93	−49.8	
5	British Steel	181	52	−71.5	Now private
6	BL (British Leyland)	176	Taken over, now private		
7	Electricity Council etc.	158	121	−23.7	Now private
8	General Electric	155	105	−32.3	
9	ICI	148	118	−20.6	Since split
10	Lonrho	140	106	−24.1	
11	Courtaulds	115	45	−61.0	
12	British Petroleum	113	106	−6.5	
13	Grand Metropolitan Hotels	109	82	−24.1	
14	Guest, Keen & Nettlefolds	107	30	−82.5	
	Total (excluding item 6)	2,363	1,447	−38.8	

Note: For the period 1976–81, the largest redundancies of these corporations and their regional incidence were reported in Townsend, 1983: 74, Table 4.1
Source: *The Times 1000, 1980–81*, 1994 London: Times Books

leading employers. Table 4.3 is based entirely on published figures (precise comparison between different years is almost impossible to attain just because of the constant process of acquisition, merger, privatisation and more recently de-merger). Some of the heaviest job losses are seen to lie in the coal and steel industries (both still in government hands at the beginning of the period) and in the restructuring of two manufacturing corporations that were very prominent in the geography of the 1979/82 recession – Courtaulds, in textiles, and Guest, Keen and Nettlefolds, in metal-using industries (Townsend, 1983). Job losses were small only in the oil TNCs and in telecommunications, the postal service and one hotel and catering chain. (As a result of these changes, the leading employers of the UK now include more banks and retail chains.)

The scale of downsizing can be estimated by summing the table (necessarily omitting British Leyland), *producing a reduction in global employment of 38.8 per cent, similar to the change for 'production industries' in all ownerships at home.* What appears to have happened then is that the relative growth of some of these corporations abroad hides an even worse performance than the rest of industry, in a period equally notable for a relative increase in the importance of smaller plants. For example, in the period from 1971 to 1988, the number of manufacturing establishments employing 10–19 people increased by 3.2 per cent, whereas those employing over 1,000 fell by no less than 60.0 per cent. In 1991, some 55 per cent of those employed in EU 'enterprises' in the

EU area worked in firms with less than 100 employees, almost 30 per cent of those with under ten.

THE EU RESPONSE IN TERMS OF INDUSTRIAL AND MERGERS POLICY

There are safeguards from the EU by which takeovers and mergers may be controlled. However, trade unionists and others would be aware of how much the European Commission is subject to lobbying by industry and is prey to the interests of national governments. Thus the Commission abandoned prescriptive proposals in 1991 after these failed to win the support of member states anxious to protect their own national system. A European directive of February 1996 requires member states' own rules to ensure that the company to whom the bid is addressed must have enough time and all the necessary information to allow it to take a properly informed decision on the bid. It also aims to ensure equal treatment for all shareholders. In fact, however, the new slimmed-down proposals involved little or no change, in most cases, to national rules and tradition. Later in 1996 it put forward proposals for powers to investigate more mergers, because too many with cross-border impact fall outside the Commission's remit (*Financial Times*, 10 July, 1996: 2).

DEINDUSTRIALISATION

Enough has been said to show that international investments commonly fall prey to the general trend of job loss in industrialised countries; foreign firms, as in the UK case, may at times lose jobs more slowly than their domestic counterparts, but the effect of TNC mergers may be sharp and adverse for individual areas, and investment abroad may be connected with a run-down of home employment. I will now turn from a collective view of firms' behaviour to their overall part in changing regional employment structures.

The decline of the large industrial corporation and its larger establishments brings into question whether we can still speak of 'industrial' countries, regions or sub-regions. Perhaps 'industrialised', in the past tense, 'once-industrialised' or 'ex-industrialised' would be a more truthful categorisation, not only for former coalfield areas of north-western Europe but even for areas of twentieth-century industry around London (or US counterparts).

Again, however, Southern Europe shows a more distinctive pattern. It was here that it was possible to find areas of industrial job growth in the last twenty years, the last occasion being the years 1985 to 1990, notably through the later investment by TNCs in new motor manufacturing plants near Lisbon and Valencia. The cyclical upswing in EU manufacturing employment from 1985 to 1990 registered itself more firmly in Iberia than elsewhere, where it represented still only a weak or minor interruption in the overall process of job reduction.

Definitions of deindustrialisation

While cycles of structural decline and resource depletion could always be found in economic history to cause depression in particular districts, it was the global element of the inter-war depression that first etched a map of widespread closures and unemployment across the industrial geographies of the richer countries. New production then found its own regions for development, but tended in its post-war peak of post-Fordist investment to overflow back to less occupied regions. It was only with the return of less profitable conditions in the 1970s that a more fundamental problem was seen:

> Within a country it was not surprising that a region which had begun to decline should continue to do so; cumulative processes were clearly set up which tended to perpetuate the decline (unless strong policies were adopted). There could be much the same sort of decline for individual countries in a free trading world.
>
> (Blackaby, 1979: 266)

Also writing on 'deindustrialisation', Singh (1977) argued that the defining characteristic was a manufacturing sector that failed to sell enough of its products abroad to pay for the nation's import requirements. We can identify several more elements.

A first critical feature of deindustrialisation, hastened by global competition and corporations investing abroad, lay in manufacturing's declining share in all the big economies. It now accounts for only 23 per cent of the US GDP, and slightly less in the UK. Even in Japan and Germany, the strongholds of industry, it accounts for no more than 30 per cent of GDP. The position is, of course, even more extreme at the level of individual regions. Of 54 regions of E12 for which the equivalent figure was known for 1990, eighteen derived less than 30 per cent of GDP from industry.

The second feature of deindustrialisation, after the loss of production abroad and to other sectors, comprises a disproportionate loss of employment and income, seen in the precedent of the USA.

> All Americans used to be in roughly the same economic boat. Most rose and fell together, as the corporations in which they were employed, the industries comprising such corporations, and the national economy as a whole became more productive – or languished. But national borders no longer define our economic fates. We are now

in different boats, one sinking rapidly, one sinking more slowly, and the third rising steadily. The boat containing routine producers is sinking rapidly.

(Reich, 1991: 208–9)

Production tasks have shown themselves amenable to successive rounds of mechanisation and automation. However, redundancies in nearly all manufacturing industries in the UK recession of 1979–82, which devastated many industrial towns through eliminating obsolete products and overmanning (Townsend, 1983), have still not been replaced by other jobs, while East Germany has now gone through the same experience, and recession conditions across much of the EU in the early 1990s have produced versions of the same pattern.

In the UK, the economic recovery of the years 1982 to 1990 failed to show anything but a minute and temporary increase of manufacturing jobs in a small minority of areas, sub-sectors and years. This repeated type of performance has earned Britain a suspicion for continuing fundamental faults in its political economy (Martin and Rowthorn, 1986). 'There can, however, be little doubt that, by promoting structural interdependence between national markets, [TNCs] have made the UK more vulnerable to changes in international demand and supply conditions' (Dunning, 1988: 195–6).

WHERE AND WHEN FACTORY JOBS PASSED THEIR PEAK

On reflection, however, the UK is only the worst case among many of de-industrialisation. Most of the other older industrial countries of north-west Europe passed their peak numbers of factory jobs in the mid-1960s, and only Japan stood at an all-time high in the 1992 data of OECD (1994a). It is the East Asian newly-industrialising countries that represent the main world group where the proportion of total jobs that are in industry is still increasing.

Employment trends

To recap on the general model (Chapter 1), we expect and find that the proportion of employment in different countries found in primary industries tends to reduce heavily with higher per capita income. In European experience the balance of workers flows in roughly equal measure to the industrial and service sectors. However, increased industrialisation eventually gives way to a divergence between industrial and service employment, with a falling off in industrial employment to the benefit, at the highest income levels, of the service sector, entering the 'post-industrial society'. These changes in the composition of economic activity are of great significance for the study of space and place because each of the three main sectors has characteristically different location patterns.

What is remarkable is that the number of actual jobs occupied (in industry and elsewhere) increased only slowly in the EU from 1975 to 1985. Much of the rise in unemployment can be attributed to slow growth or actual decline in the numbers at work in E12 industry, which changed from 48.4 million in 1968 to 47.0 million in 1980 and 39.8 million in 1992. Here there were permanent reductions in factory employment, notably in France, The Netherlands, Germany and Italy, together with Britain. There was to some extent a process of dispersal at work which spread jobs to Southern Europe, before, of course, the 'new international division of labour' took firms further afield to Eastern Europe and beyond. The EU recognised 'outward processing' as a regular feature of its industrial activity. However, this 'pull' of cheaper labour from the periphery was only accentuating a trend under which the 'push' of high labour costs and dramatic effects of technological change tended to encourage automation and redundancy in the industrialised core of Europe.

The historical peak of industrial employment (absolute numbers)

Looking back at the process of European deindustrialisation, it is remarkable how long ago the process started, if from OECD (1994a) data we establish the peak year of each individual country's profile of industrial employment. The earliest countries to peak were a group that had followed Great Britain in the very process of industrialisation itself, led by Belgium, Germany, The Netherlands, Switzerland and Sweden in 1964 to 1965, although Britain itself did not pass its highest total of industrial employment until 1966. So there is clearly both an association with early heavy industry on the one hand, and with wealthy progressive societies on the other, in the early start of deindustrialisation.

At the same time, it is quite common to find authorities (e.g. Dicken, 1992) referring to Southern and Eastern Europe among the newly-industrialising countries, even perhaps including Italy by virtue of the 'Italian miracle' of 1958–64; accordingly Italy had a later peak, in 1980, while the other European countries attaining their maximum level in the 1980s – Ireland, Finland and Iceland – were small and peripheral. This left only Portugal and Greece to establish their first major downturn in industrial employment at the beginning of the 1990s, the peak year for the EU as a whole having been passed long ago in 1970, and for western Europe as a whole in 1974.

The historical peak of industrial employment (relative proportions)

Relative 'deindustrialisation' is, however, the more meaningful measure, incorporating the greater growth of population and of associated service jobs. In terms of absolute numbers of jobs, the USA and Canada showed their highest figures in 1989. If, however, we take industrial employment as a proportion of total, we find that the peak year of the USA and Canada for dependence on industrial work occurred earlier, indeed before 1960, and in 1973 in Japan; this definition is more relevant in these non-European economies than the absolute disappearance of industrial jobs, which in the USA was more specific to particular times and places.

Figure 4.3 reviews the proportionate size of industrial job losses for leading countries of OECD and identifies the year in which industry's share of total employment in each country attained its peak. Because of the mounting growth of service employment in the 1960s, this date was generally somewhat in advance of the absolute peak of industrial jobs as described above. Thus, the UK showed its relative peak in 1960, a full six years before the absolute one, and was in fact the first European country to do so. Most of the northern European countries had followed by 1970. However, two countries that still have over a third of their workforce in industry, (West) Germany and Austria, did not peak until 1970 and 1973 respectively, while the four countries of Southern Europe passed their peak only in the years 1971 to 1982. Having had more time to 'deindustrialise', the countries of northern Europe have shown much greater reductions in industry's share of total employment; the graph clearly gives the UK the biggest reduction, from 47.7 per cent to 26.5 (on a wider definition than manufacturing alone).

This diversification into service jobs is also very relevant to the balance of advantage between men and women; for example, men's jobs reached their peak level in the UK in 1970. The overall pattern of industry's declining share of jobs in OECD countries is summed up in Figure 4.4, and graphed

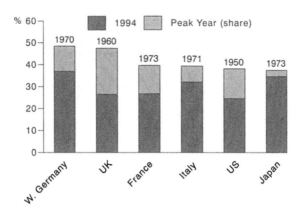

Figure 4.3 Decline in the industrial share of total employment from its peak year (named) to 1994
Note: Last data for USA and Japan are for 1995
Source: CEC, Eurostat, 1996c; OECD, 1995

in simple fashion against men's declining share of total employment, on the horizontal axis. On this basis, deindustrialisation appears to have proceeded furthest in Norway and Australia, whereas men's share of total jobs has fallen nearest to 50 per cent in Finland and Sweden. The precise trajectory taken by individual countries varies, but the identification of men's work with a declining sector carries many implications which I will explore further in Chapter 6.

CHANGES IN THE NATURE OF DEINDUSTRIALISATION?

The impact of deindustrialisation in successive recessions has now merged to create the EU's international plateau of high unemployment. The early post-war establishment of the European Coal and Steel Community pre-figured the heavy structural problems in areas with those specialisms. Beyond that, older and heavier types of engineering (including shipbuild-ing) tended to contribute more to structural decline, and job loss advanced in lower technology industries generally. By the period 1970 to 1990, how-ever, Europe saw fairly even decline across different countries in different manufacturing industries, although with markedly more countries suffering decline in textiles and markedly fewer in paper and chemicals. Taking the

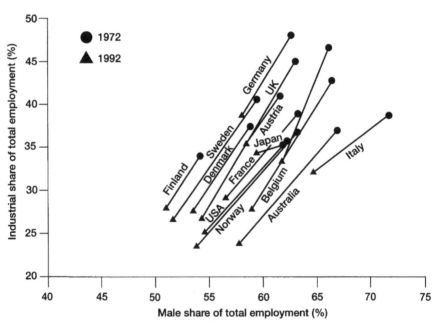

Figure 4.4 International correspondence between the decline of the industrial and the male shares of total employment
Source: OECD, 1994a

84

longest period of data available for the E10, from 1983 to 1991, we see that job losses were concentrated in energy industries, metal manufacturing and chemicals industries, with further losses in textiles, clothing and footwear.

Privatisation

One feature that was increasingly exported as a policy from the UK to the rest of Europe, East and West, was that of privatisation – the sale of nationalised industries and utilities to private sector bidders. This is relevant to the theme of our chapter in three ways. In the first place, there has been a tendency for governments to place loss-making but important industries under public ownership, supported by subsidies. Secondly, it can be argued that trade union political influence had the effect of restricting job losses in state monopolies, and that the essence of the process of privatisation, therefore, was to generate profit through the slimming down of workforces both before and after the formal transfer to new management (Table 4.3, above). Thirdly, there was a tendency for this type of activity to be located in more industrial areas (Champion and Townsend, 1990). Job losses continued in 1992–6, with reports affecting the steel, electricity, gas, water and telecommunications industries, with tens of thousands of job losses forecast in gas and the railways – fears that were better grounded than in health and education.

Defence cuts

Global change brought defence cuts as a later addition to the pattern of industrial and geographical job loss, indeed one that added a major new element to the balance of regional economies. According to the Ethical Investment Research Service (EIRIS, 1990), more than half of Britain's top fifty companies made money from defence contracts, while union researchers then estimated that 11 per cent of industrial production was defence related and that a third of jobs in defence manufacturing would be at risk. Despite some attempts at job diversification for defence factories, there have been cuts in British Aerospace's military workforce from 48,000 to 31,000 over five years, bringing experience of heavy redundancy for the first time to counties such as Surrey and Hertfordshire, additional to many job losses round the coast at shipyards and naval bases.

In continental Europe, by contrast, entrenched employment rights and a political climate that opposes large-scale redundancies mean that companies there remain heavily overstaffed. They are therefore vulnerable to US competition, given that the industry in the US has cut its costs through a mixture of rationalisation, redundancy and corporate restructuring; in the US, 'the recent debate over the peace dividend suggests that politicians are acutely aware of the potential gains and losses of the end of the Cold War for their local economies' (Trubowitz and Roberts, 1992).

The future

It would be wise to note that the worst effects may be yet to come. Although Wilson (1994) expected job losses from the primary and manufacturing sectors to be much more modest than in the past, this has already been belied by the outcome of the recent recession. Reich (1991) persists with the view that the numbers of US production jobs will continue to 'sink fast', while there are suggestions that manufacturing might fall as low as 10 per cent of total employment in the next thirty years, pointing us to our next chapter:

> Manufacturing companies in the west will face competing options. They can continue to compete on cost if they switch production to developing countries. Or they can seek high value-added niches with a big service component. Either way there will be a shift of employment to services
>
> (Brittan, 1993: 22)

SUMMARY

- 'Global shift' is not purely and simply the main cause of 'deindustrialisation', for the FDI investment of TNCs in rich countries is several times more important than in poor.
- TNCs' command varies greatly by sector; however, with their big role in trade and in running large plants, TNCs do impart simultaneously elements of vulnerability and of opportunity to individual areas, and do contribute to general job loss.
- Indigenous small firms are now doing better in job creation terms, while the TNCs can be said to have contributed disproportionately to the statistics of deindustrialisation.
- These show a progressive spread of net job loss from the oldest industrial countries of the EU to the newest.
- Deindustrialisation takes on new forms through the international spread of privatisation and cuts in the manufacture of defence equipment.

5

SERVICES – THE 'AMERICANISATION' OF EUROPEAN JOB STRUCTURES?

'Relative deindustrialisation', taken as industry lessening its share of total jobs, started first in North America (Chapter 4). Here, large volumes of new jobs outside industry were accompanied by the growth of low paid work, especially in women's, part-time and service employment. I now ask whether the much criticised direction of North American employment change is being followed in western Europe.

The chapter proceeds to identify key characteristics of the US labour market, establish contrasts and similarities between the US and western Europe and explain the role of the service sector in employment trends. This chapter's study of the western European pattern shows some variety between countries in the recent strong trends for growth in women's, part-time and service employment, with fundamental effects on the daily working of society. These trends were accompanied in the USA by growth in new areas and by decentralisation of services from metropolitan to out-of-town and rural locations, a pattern we consider for western Europe.

INTRODUCTION

In assessing the future of jobs in EU countries, it is useful, as in Chapters 2 and 4, to make international comparisons, and to ask two question in particular. Does North America act as a warning to Europe as regards its future socio-economic condition? And are the Western industrialised countries each following similar paths to new employment structures?

There are some views that the USA and the EU countries are following *divergent* paths. For instance, rates of unemployment, which in the 1970s were appreciably higher in the USA than in Europe, have been firmly overtaken by EU countries in the last ten years (Figure 2.4a). Lipietz (1993) stresses a distinction of process, arguing that while US expansion in the 1980s was based on flexibilisation and de-skilling of productive processes (as defined in previous chapters), leading European countries, notably France and Germany, took the alternative track of specialisation, concentrating on skilled specialised work; this pattern did not, however, extend Europe-wide, for the UK and Spain, he argued, adhered more to the US track. However, the contrast points us towards widely observed and crucial concerns over American capitalism itself:

1 A first feature is the further reduction of manufacturing's share of GDP;
2 A second feature involves productivity, skills and the distribution of earnings, with a heavy slowing of the US historic rate of growth in productivity being associated with mounting social inequality;
3 This is involved with a US business culture which, among other things, generally provides worse conditions of employment than in the EU, notably for women.

Historical background and approach

There is little new in the growth of services. As early as 1900 the USA and Britain had more jobs in services than in industry. By 1950, services employed half of US workers (Figure 5.1) and the industrial share of total jobs entered a steep and continuous decline. Services have since increased to account for more than 70 per cent of both employment and GDP. The principal argument considered here is that the US swing towards a service-

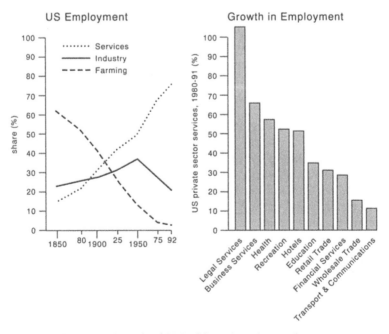

Figure 5.1 Growth of United States' service employment
Source: *The Economist* 20 February 1993, from US Labor Statistics Bureau

88

based economy is merely the earliest and largest example of a general pattern of change in developed countries.

Originating in papers of 1959–62, Bell's *Coming of Post-Industrial Society* (1974) argued that a great divide began to open up from about 1947, with the emergence of a white collar society and of far-reaching processes of change centred on the role of science and technology, the increased role of the professional and technical classes and the growth of the service sector of the economy, particularly in health, education, research and government. The processes of change were merely more advanced and visible in the USA. It was possible to speak of international convergence in the arrival of post-industrial conditions:

> If there is a meaning to the idea of convergence, it is that societies resemble each other somewhat along the same dimensions, or they may confront a common core of problems. But this in no way guarantees a common or like response.
>
> (Bell, 1974: 114)

It is in this spirit that this chapter adopts an international form of analysis, allowing that countries do vary in their speed and type of response. Although Germany, Austria and Switzerland and, to a greater extent, Southern Europe, remain 'behind' North America in their tertiarisation and employment of women, nonetheless I shall argue here that the same forces and processes are imposing similar directions and even rates of change on the labour market, and ultimately on urban and regional change, only with a timelag. This is not a single process, but a set of comparable trends.

US 'JUNK' JOBS?

Titles such as *The Age of Diminished Expectations* (Krugman, 1990), *The Politics of Rich and Poor: Wealth and the Electorate in the Reagan Aftermath* (Phillips, 1990) or *America: What Went Wrong?* (Barlett and Steele, 1992), which were not widely available in Europe, attest to a widespread reaction against the nature of the US economy. Recent years have shown strong awareness of the growth of US jobs with poor pay and conditions:

> Controlling for family size, geography, and other changes, the best estimate is that between 1977 and 1990 the average income of the poorest fifth of Americans declined by about 5 per cent, while the richest fifth became about 9 per cent wealthier.
>
> (Reich, 1991: 197)

It is argued that no one understands the reason for increased dispersal of income in the 1980s (Krugman, 1990). Real earnings in manufacturing have been nearly static over twenty years. However, as services were the main area of change in the origin of earnings in this period, this polarisation of income levels, without much improvement in the average, can be attributed to that sector.

Thus, although there was an increase of 43.1 per cent (massive by European standards) in civilian employment from 1972 to 1992, it was accompanied by surprisingly unfavourable features. As real hourly payment failed to increase over the 1980s, the growth of employment income was sustained only by rising economic activity rates, at the cost of either leisure or of family and domestic life (US holiday entitlements and days off work with sickness stand at less than half the levels of the UK, France or Germany, and hours of work remain much greater). The failure of men's incomes to rise was offset only by the greater participation of women in paid work. Over 78 per cent of adults of 16 and over are in the US labour force, compared with 58.7 per cent in E15 in 1994 (CEC, 1995b).

In effect then the US had switched quite suddenly from an economy based on high levels of income and productivity growth toward one of work-spreading. OECD Economic Surveys (1992) argue that while real wages and salaries failed to rise, US productivity did increase, but disappointingly, by only 0.8 per cent per year in the 1980s. To some extent it may reflect the increasing share of the services sector, where productivity and productivity growth traditionally have been lower and harder to measure.

The 'junk job' is not surprisingly a stereotype for young workers, part-time students, and many groups of women and ethnic staff, and is a cause of lower income and social exclusion at the bottom of the income distribution. Nonetheless, we accepted talking above about 'polarisation' toward *two* extremes. There are several ways of appreciating the scale of job growth in lucrative service industries, such as law and business services (Figure 5.1). Esping-Andersen establishes statistical totals for 'junk-jobs' and 'post-industrial jobs' and concludes:

> If America's large 'junk-job' component supports a proletarianization thesis, other factors point in the opposite direction. First, over the long haul, professional jobs grew faster than bad jobs in the United States. Second, the middle group of service jobs is much weightier than the bottom group.
>
> (Esping-Andersen, 1990: 208)

Or again, contrary to popular belief, the 'new service jobs were not mainly low-paying jobs of the hamburger-flipping variety. America created proportionately twice as many high-skilled jobs as France and Western Germany. Indeed, four-fifths of all new jobs created in America were in the professional, technical, administrative or managerial category' (*The Economist*, 19 November 1994: 100).

Exploited women?

'The fact is that a service economy is very largely a female-centred one' (Bell, 1974: 221) and therefore especially connected with low pay in the USA. The growth of services in the US economy is strongly dependent on the increased use of women. Over the period 1972 to 1992, women's employment increased by 72.1 per cent, and services by 64.6 per cent, to a position where women made up 45.7 per cent of the labour force in 1992. These increases outran all forecasts and reflect the social revolution that we see in American media, with increased participation by younger married women, and more working mothers.

However, viewed in economic terms, women's great increases in education and experience led to only slight reductions in occupational segregation and to only narrow groups achieving better income parity with men. Overall, women's median annual earnings fluctuated around the level of 60 per cent of men's, low by European standards. Smith and Ward (1980) investigated the question of why this ratio remained constant at the same time that women's participation in the labour force expanded rapidly. These conditions and others weighed heavily against large gains for any group. Industries that absorbed the largest influx of women workers – those in the rapidly expanding service sector – were those with traditionally low hourly earnings (Bianchi and Spain, 1986). Possibly, women took work at a lower price because men's wages were also stagnating, but beyond all that it is clear that it suited capitalism to segment the labour market.

Remembering our question about 'convergence', it would seem that the exceptionally rapid growth of women's jobs will in future be moderated, for demographic reasons and because of the limited number of non-working women left.

CONTRASTS IN POLITICAL ECONOMY

All this occurred in the context of an ever-weakening role for organised trades unions in a highly organised, business-dominated culture in the US (Sexton, 1991). Some hold that job growth and social protection are inversely related. The trade-off between job growth and social protection is different in the two continents, with the US biased toward business growth and the EU countries more concerned with social protection. The trade-off also results in different distributions of income; Figure 5.2 shows the earnings of workers in the highest and lowest deciles (tenths) of the income distribution as a proportion of median earnings. A European in the lowest decile earns 68 per cent of median European pay: his/her American counterpart earns just 38 per cent of European pay. The risk of poverty is worsened by the threat of unemployment, and the harsh limitations that apply to payments to the unemployed.

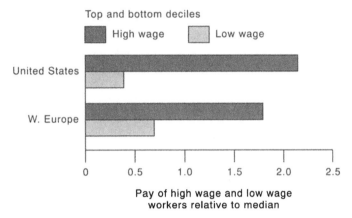

Figure 5.2 The greater extremes in the US income distribution
Note: W. Europe here comprises Austria, Belgium, Denmark, France, Germany, Italy,
The Netherlands, Norway, Portugal, Sweden and the UK
Source: *The Economist* 12 March 1994 (after R. Freeman, OECD)

In Europe the Kreisky Commission on Employment Issues (1989) there-fore painted a sombre picture of American income structures and poverty:

> The results of the Harvard study raise serious doubts as to the suit-ability for Europe of the American model of high labour market flexi-bility. The 'working poor' in the United States are the equivalent to the unemployed in Europe. A participant of one of our meetings argued that he would rather be unemployed in Germany than 'working poor' in the United States.
>
> (Kreisky Commission, 1989: 148–9)

The Commission did recommend to the CEC certain manpower strategies involving greater use of flexibility and, above all, a reduction of working hours. This accords with views of Lipietz (1992) that in France a cut from thirty-nine to thirty-five hours per week would create a million extra jobs over five years, and agreed with widespread views in France, Italy and Germany of the merits of 'work-sharing'.

ARE WESTERN EUROPEAN EMPLOYMENT STRUCTURES FOLLOWING THOSE OF NORTH AMERICA?

Western Europe, despite its best intentions and policies, may already be following the USA in its internal economic structure, as well as interacting with the US in trade and through TNCs' interpenetration of investment and acquisitions. A more careful examination of the dominant secular trends of employment might reveal that, in certain crucial aspects, western Europe is following North America into a marked dependence on women's and service employment.

Was the USA ahead of Europe in women's employment?

There must of course be reservations to our main hypothesis. For some countries, it is actually debatable whether there was any increase in service and women's *full-time* employment in the hundred years prior to the 1960s. Following from that, it is not entirely obvious that major changes toward a post-industrial economy commenced earlier in the USA than in any European country, although the last figure (5.1) did date the onset of relative industrial employment decline for the US back in the early 1950s. The most consistent international data source (OECD, 1961) shows how by 1956 the US already had 54.3 per cent of civilian employment in the service sector, followed by Canada with 47.3 and the UK with 46.4 per cent; at that time, Italy was recorded as having only 28.3 per cent of workers in services.

However, the US was then by no means exceptional in the development of women's work, with 32.0 per cent of its civilian workforce being women. Christopherson (1989) attributed this to the dominance of men in the negotiation of work contracts in mass production. The US proportion was indeed exceeded, notably in Sweden and Denmark, as in Japan and Germany, where the proportion of women recorded in the total of employment was still tending to *decline* as part of reductions in the recorded role of unpaid family labour in agriculture.

All in all, however, the rapid uptake of women in the USA in the 1960s made it a leader of the field in terms of the crude numbers of women at work. Today women's share of total employment in the US is exceeded only in Scandinavia and equalled in the UK.

Twenty years of parallel change

There are many sensitive social, cultural and political factors involved in the differentiation of women's participation rates in, say, Scandinavia from other areas. Esping-Andersen (1990: 208) suggests that 'national variations are tremendous, and that a generalization from the American experience is unwarranted'; for example, 'the Swedish experience shows that post-industrialization can be very strongly biased in favour of professionals', with the role of the welfare state being one of the critical differentiating features. For instance, Sweden has enjoyed a strong dynamic from welfare state recruitment of women, and state support for mothers in work: there are very few 'junk jobs'. Germany has a more generally segmented labour market, with less 'junk jobs' and polarisation than the USA. Bell (1974) agreed certainly that any question of convergence of national societies was debatable.

However, whatever the different 'starting levels' of different countries in their relative participation rates, these countries tend to retain the same rank order position over time as they move forward their dependence on women's and service employment at parallel rates; thus, the USA and Scandinavia shared the lead in women's employment both in 1956 and 1992, over a quarter of a century, and jointly led other countries.

Taking one dimension to test the alternative ideas of parallel change, convergence or divergence, Table 5.1 outlines the principal statistical changes reflecting women's employment over twenty years, 1972 to 1992 (but see Chapter 6 for a qualitative view of less visible 'informal' work). The USA showed a massive increase of 22.5 million (72.1 per cent) in its number of women in civilian employment and increased their share of total employment by 7.7 percentage points to nearly half the workforce. While Europe (see E12) started behind the USA in 1972, and increased its absolute number of women's jobs at half the rate, perhaps due to a much less dynamic growth of population, nonetheless western Europe as a whole (E12) showed an increase of 14.4 million, an increase of 8.1 percentage points in the share of

Table 5.1 Female employment growth and shares of total civilian employment

Country	Thousands		% share of total jobs		
	1972	1992	1972	1992	Change in % points
United States	31,257	53,793	38.0	45.7	7.7
Canada	2,885	5,568	34.6	45.5	10.9
Australia	1,852	3,246	33.0	42.3	9.3
Japan	19,570	26,190	38.2	40.7	2.5
Denmark	965	1,217	41.0	46.6	5.6
Finland	965	1,060	45.8	49.0	3.2
Norway	601	913	36.6	46.3	9.8
Sweden	1,615	2,123	40.6	48.5	7.9
Austria	1,153	1,474	38.5	41.6	3.1
Belgium	1,214	1,535	33.5	41.2	7.7
France	7,555	9,635	36.7	43.7	7.0
W. Germany	9,848	12,069	37.4	42.0	4.6
Ireland	279	377	26.8	33.9 ('91)	7.1
Switzerland	1,102	1,367	34.0	38.2	4.2
United Kingdom	9,041	12,136	36.9	45.7	8.8
Greece	863 ('71)	1,310 ('91)	27.5	35.2	7.7
Italy	5,308	7,587	28.2	35.7	7.5
E12	40,579	55,078	33.7	41.8	8.1
OECD area*	101,945	151,470	36.0	43.8	7.8

Note: 'OECD area' represents virtually all 'developed' countries; (*) indicates that its first 'developing' country member, Turkey, is excluded from the table; data for 'E12' and 'OECD area' include estimates for component countries that are otherwise excluded from the table because of significant discontinuities in the source data.
Source: OECD, 1994a

all work (slightly greater than the US). Clearly, there is lower equivalent growth recorded in the table for a miscellany of countries, Japan, Austria and Finland.

The argument here is essentially, however, that almost all the countries, many with different value and welfare systems from the USA, are following the same kind of trajectory. Across the OECD area (excluding Turkey) growth in women's employment reached 48.5 per cent, and easily withstood the impact of recession (Paukert, 1984), continuing in the early 1990s recession in Britain (Townsend, 1994a). Clearly, the Scandinavian countries, which like the USA and Canada had higher purchasing power than the E12 countries, have tended to run ahead and Southern Europe (Greece, Italy, Portugal and Spain), to run behind most of the EU. However, women's recorded share of total jobs increased at approximately similar rates throughout the EU, not least due to the systematic effect of decline in men's manufacturing jobs.

The answer to the question of whether Europe is following the North American pattern of evolution in its labour markets is therefore that, from different starting positions, most countries are moving at about the same rate towards a post-industrial economy. The USA is not the absolute leader in regard to women's employment patterns because its statistical position has been shared by the UK and preceded by most of Scandinavia, with its emphasis on welfare state recruitment in the period under study. These statistical generalisations do not imply social convergence.

PART-TIME VERSUS FULL-TIME WORK

The growth of part-time work may also give an exaggerated idea of women's gains in total paid work and income. Two major features distinguish men's and women's employment trajectories: women are much less likely to have continuous occupational careers and they are more likely to work part time, and these differences are primarily a direct consequence of the reality of motherhood and society's treatment of it. The extent of part-time working is negligible amongst men and a significant, albeit minority, practice amongst women. Thus data for E15 (CEC, 1995b) shows an increase in men's part-time work from a share of 3.3 per cent of all employment in 1985 to 4.8 in 1994, whereas women's rose from 27.2 to 30.3 per cent.

These are moderate figures, so Denmark, Sweden, Norway, the UK, and above all, The Netherlands are outliers from the rest of E15 in showing a high dependence on part-time work (Figure 5.3). Using a system of full-time equivalents (Townsend, 1986) women employees' work in the UK expanded by only 15.7 per cent from 1971 to 1996, and it would appear that Table 5.1 does overstate the size and growth of women's work, at least more than it does for other countries. The USA actually has lower proportions of part-time workers than the UK and The Netherlands; indeed for various reasons a

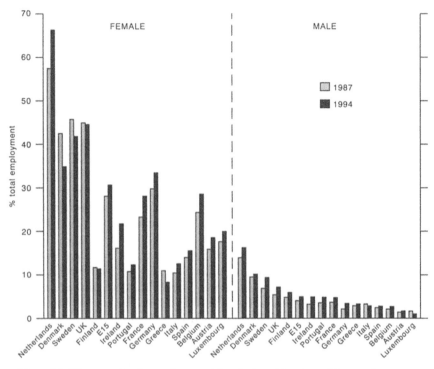

Figure 5.3 The growth of part-time working (below thirty hours) in member states, E15, 1987–94
Source: CEC, 1995b

trend away from part-time work was reported in the late 1980s and 1990s (Christopherson, 1996). In comparing experiences of women in Britain and the USA, Dex and Shaw (1986) found that British women were more likely to return to work part time after childbirth than were their American counterparts. Dex and Shaw saw the reason for this lying partly in women's preferences, but also in the legal advantages in the UK of part-time contracts for employers.

The Netherlands exceeds the UK for its proportion of part-time women workers, around 60 per cent on certain definitions. Part-time work has greatly expanded the number of people in employment, by approximately 300,000 between 1979 and 1990. Although the early 1990s recession led to a decline in full-time equivalent employment, the number of people in work did not fall because of the continued growth of part-time working (CEC, 1994a), a pattern similar to the UK. The historical explanation for the growth of part-time working lies in the response to the excess demand for labour in the 1960s. Up to then participation of women was very low compared to other countries, and employers were forced to offer jobs to women that met their view of family responsibilities in the context of limited child-care facilities. From then on, part-time work became an accepted feature of the labour market and developed from providing second-class jobs for married women to fully acceptable jobs for all women and increasing numbers of men (CEC, 1994a).

What we are finding is that, while the strong element of part-time growth has exaggerated the strength of The Netherlands and UK women's growth of jobs, it is a somewhat irregular and sometimes quite small component of a more generally regular trend of women's employment growth at large. Indeed, the very definition of part-time employment is itself varied and irregular. The CEC (1994a) attempt to resolve the issue by ignoring countries' own classifications of who is self-employed and simply counting workers undertaking between ten and twenty-nine hours per week. When this is done a significant part of the apparent growth of part-time working (though not in the UK) seems to have been due to more jobs being classified as part-time rather than full-time rather than to an increase in the actual number of people with part-time hours.

Comparing the years 1987 and 1994, the proportion of women working these part-time hours increased virtually everywhere except Greece (Figure 5.3). The trend then is toward part-time work, but it is generally not as consistent as the wider one towards women's jobs. Although the part-time performance has the potential to distort and exaggerate our interpretation of international trends, in practice the use of full-time equivalents has the effect only of moderating the shape of analyses.

THE QUESTION OF EARNINGS

Does the EU exploit women workers as much as the USA? It is notoriously difficult to compare the earnings of men and women because of the important structural differences that persist between the worlds of work of men and women. It is even more difficult to extend the comparison across countries. However, it is clear that in the USA 'women are paid about 60 cents for every dollar paid to men' (Bianchi and Spain, 1986: 171) but that the ratio varies upwards from about 59 to 85 per cent in the countries of the EU.

Taking full-time and excluding seasonal US workers, then data of Bianchi and Spain show considerable stability in the earnings ratio of men and women, varying only between 57 and 64 per cent in the twenty-eight years from 1955 to 1983. Data are also available across all industries and sectors for Great Britain, which was unusual in adopting the Equal Pay Act, 1970. This brought the *hourly* earnings of full-time women up to 72.8 per cent of men's by 1981 and 80.0 per cent by 1994; but because of women's shorter average hours their *weekly* earnings rose only from 66.1 to 72.2 per cent between the same dates, including a significant *acceleration* over the recession years after 1990. If we look at all sources of income then part-time work (which is not counted above) is a significant addition:

> To sum up, two decades ago one pound in every four that came into UK private households went into the hands of a woman. In 1991 that proportion had risen to one pound in every three. The single most important reason for this trend has been the growth in female part-time work. Women are still heavily over-represented in the bottom deciles (tenth) when individuals are ranked by their individual income, but this situation has improved significantly in recent years.
>
> (Webb, 1994: 110)

Since the data available for the EU is restricted to average earnings of manual workers in manufacturing, it must be treated with a good deal of caution. Over the past twenty years in the EU there appears to have been some narrowing in the differential between wages of men and women, with some assistance from CEC directives since 1975. However, the relative earnings of men and women appeared to change very little between 1985 and 1991 (CEC, 1994a). Women's relative income varied a good deal across the EU in 1991, when Denmark had the highest ratio at 85 per cent and Ireland had only 59 per cent. These differences are clearly inconsistent with EU policy and a wide variety of explanations may exist, according to Duncan (1994). Equal pay policies have fallen prey to demands for moderation in pay bargaining and to the decentralisation of negotiations to the level of individual firms, which also detracts from the imposition of higher national standards.

'Most fundamentally, however, the persistence of horizontal and vertical segregation means that women are doing different jobs from men, and traditionally male skills are still more highly regarded than female ones' (Perrons, 1994: 1209). From this review of the limited data on women's earnings it appears that they are subject to many of the same restraints as in the USA. However, halting and varied progress in the direction of gender equality is evident in the EU as a result of public policy, but the underlying element shared with the US may be the need for service work at inherently low productivity levels.

THE LINK BETWEEN WOMEN AND SERVICES

Women's supply of labour

'The nexus of government and services has been decisive for women's employment opportunities over the past two decades. But it differs from country to country' (Esping-Andersen, 1990: 201). We can now link together more firmly the two prominent features of North America with which we began this chapter – the growth of women's jobs and the demand for service workers.

In assessing *how* women's employment has grown, Kiernan (1992) stresses supply-side factors, domestic relations in the family and labour-saving devices in the home. Christopherson (1989: 140) argues that an understanding of US work practices cannot be explained either by sole reference to labour demand or individual behaviour. 'An adequate explanation must take into account the politics of ethnicity and gender as well as the search for new ways to accumulate capital.'

A significant body of writers has seen the growth of women's activity rates as a function of social change. It would be difficult to deny the role of these factors entirely. However, the role of these influences can be questioned. For instance, labour-saving devices have not greatly reduced total hours of housework. And in France the post-war growth of women's employment pre-dated the arrival of labour-saving devices. Further, analysis and cross-tabulation of trends (Townsend, 1994a) serves to show that growth in the demand for women from the growth of the service sector would actually be large enough to explain women's employment trends, both internationally and in Britain.

The widespread demand for labour from services

Services' demand for labour is high in both the US and EU. Services now account for more than half the output – and well over half the jobs – in these countries. Later I shall review the growth of three leading fields of change: business or 'producer' services, retailing and tourism. Other major fields include transport and communication and a large range of public welfare services – education, medical and social services – and of personal services such as hairdressing and cleaning. Together these different realms of service account for a large part of the quality of life of an individual area (Wood, 1991b) as well as of job opportunities, and together take up a steadily rising share of countries' output and employment.

Remarkably, the composition of this growth does vary appreciably between countries and over time (Marshall and Jaeger, 1990). Thus Bell (1974) was correct in seeing the post-industrial society of the 1960s and 1970s focusing on education, health and the professions, whereas both the UK and the USA have since seen a major emphasis on the growth of producer services. Most countries have seen the mantle of growth passed on from traditional services such as transport to the social and personal and, finally, to producer services.

Clearly varied types of services tie together differently in individual countries and situations. Esping-Andersen's (1990) leading point was that nations are following distinctly different post-industrial strategies; Sweden, for example, developed a strong dynamic in the growth of good quality public service jobs for women from the development of its welfare state, contrasting with the US divergence between high and low quality service jobs and the

99

lesser growth in West Germany. As the CEC (1994a) attests, the low proportion of service jobs in Germany is not a statistical artefact.

> From the comparison of six industrialized countries – France, Hungary, the Federal Republic of Germany, Japan, Sweden and the United States – it has been shown that while the share of service activities in the total grows everywhere, this happens in very different ways.
>
> (Illeris, 1991a: 9)

There is remarkable similarity in services' pace of growth despite these differences. The overall pattern of data for services' share of employment (Table 5.2) shows a continuation of the post-1960 pattern of increase, with the USA and Canada attaining the highest levels of service employment. Bearing in mind the other countries that have passed the figure of 70.0 per cent of employment in services (The Netherlands, UK, Norway and Sweden) there is some evidence of the fulfilment of Fourastie's (1949) prediction that in advanced countries the services' shares of total employment

Table 5.2 Service employment growth and shares of total civilian employment

Country	Thousands		% share of total jobs		
	1972	1992	1972	1992	Change in % points
United States	51,787	85,260	63.0	72.5	9.5
Canada	5,237	8,932	62.8	73.0	10.2
Australia	3,165	5,492	55.8	70.8	15.0
Japan	25,110	37,980	49.0	59.0	10.0
Denmark	1,318	1,796	55.9	68.7	12.8
Finland	959	1,373	45.5	63.5	18.0
Norway	875	1,398	53.4	71.0	17.6
Sweden	2,154	2,943	55.8	70.2	14.4
Austria	1,400	2,035	45.0	57.4	12.4
Belgium	1,965	2,596	54.3	69.7	15.4
France	10,010	14,525	48.7	65.9	17.2
Germany	11,741	16,799	44.6	58.5	13.9
Netherlands	2,669	4,698	57.1	71.4	14.3
Switzerland	1,516	2,108	46.7	60.6	13.9
United Kingdom	13,107	17,947	54.3	71.3	17.0
Italy	7,792	12,670	41.3	59.6	18.3
Portugal	1,250	2,486	37.7	55.3	17.6
Spain	4,680	7,102	38.0	57.5	19.5
E12	56,173	80,619	46.6	63.5	16.9
OECD area*	149,001	229,054	52.6	66.3	13.7

Note: 'OECD area' represents virtually all 'developed' countries; (*) indicates that its first 'developing' country member, Turkey, is excluded from the table; data for 'E12' and 'OECD area' include data for component countries which are otherwise excluded from the table.
Source: OECD, 1994a

would attain this level, but his prediction was that they would 'plateau off' at around 70 per cent.

Looking at change over the period 1972–92 there is a sense in which western Europe can be seen to be 'catching up' on the rest of the 'developed' world, because the USA, Canada and also Japan each added 10 percentage points to services' share of their total jobs, whereas the E12 showed an increase of no less than 16.9 points, and the Southern European countries exceeded this.

More women service workers everywhere

Despite some variation in countries' historically rooted starting levels, there is nonetheless a strong statistical connection between the growth of women's and service employment. Its significance is strengthened by several features. Firstly, growth of services in the last twenty recorded years by 83.5 million jobs has exceeded employment growth in the OECD area as a whole, though this was marginally not the case in the USA. Secondly, the growth of women's employment in services is nearly always the largest single category of employment change, counting at the very least for a majority of job gains, as in the USA. Thirdly, these large increases may be termed 'structural'; they are achieved not principally by women taking over existing or new men's jobs, but simply by the expansion of sectors in which women were already well represented. Calculations were made for the USA, Italy and Norway over the last ten years to see what proportion of the increase in women's jobs can be accounted for simply by the disproportionate expansion of women-employing sectors. In the USA and Norway the proportion was over three-quarters, while in Italy it was well over that.

The new demand for women's labour from services, on both sides of the Atlantic, was the leading contributor to the large increases in their employment in recent decades. Of course the domestic circumstances that we mentioned earlier facilitated the change. It is suggested then that the direction of causation lies from the existence of a high income economy to an increasing need for services of the kinds that already relied on women's labour (Townsend, 1994a).

Figure 5.4 demonstrates by means of a graph the strong correlation between these major expansions of service jobs and those of women's employment (including full-timers and part-timers equally as in Table 5.1). The graph refers to individual OECD countries. The movement of an individual country is charted over a quarter of a century from the lower date, 1972, to the higher position, 1992. On the horizontal axis, the percentage of civilian employment that was female varied in 1972 between 28.2 per cent in Italy and 45.8 per cent in Finland, or 35.7 and 49.0 (no less) in 1992. The vertical axis displays the percentage of employment in services (on the same international definitions employed in tables above). There is a simultaneous

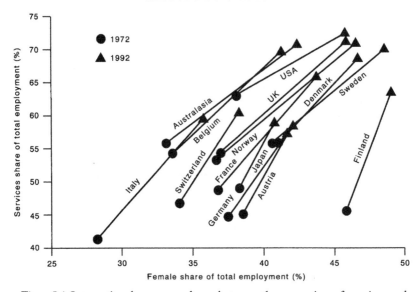

Figure 5.4 International correspondence between the expansion of services and
female shares of total employment
Source: OECD, 1994a

increase in proportions of both women's and service employment, indeed
most developed countries are moving in parallel. Clearly the swing towards
services is even more pronounced than the move towards women's employment which, I argue, it largely generates. Approaching the level of 70 per
cent in services, it seems clear that growth is especially dependent on services. Scandinavian countries are closest to having a 50/50 male/female
workforce, but at a local level, as in the UK, there are many areas with more
women employees than men.

SERVICES IN REGIONAL DEVELOPMENT

Women's employment has certainly been an element contributing to the
equalisation of employment levels between regions. The growth of women's
employment has contributed toward processes of regional 'convergence'
(the elimination of longer-standing differences in regional prosperity; Chapter 3). That is particularly through their recruitment, as welfare state provision (in women-employing sectors) has been built up to meet general
national standards in peripheral regions. Women's employment can involve
more varied kinds of change in Europe than in North America. Christopherson (1989) argues that the massive growth of service employment has
reduced differences between places and that this is of itself important for
geographers.

The most distinctive elements of economic geography are likely to remain those of the manufacturing and primary sectors. What, however, are the implications of the view that manufacturing might reduce its workforce to 10 per cent of the total in most OECD countries (Brittan, 1993)? Although producer services (Marshall *et al.*, 1988) have been added to the geography of production, and chapters 7–9 follow on individual services, it remains difficult to assess the impact of the growth of services on regional development as a whole. However, it is clear that they can bring income into an area, and that they may be the leading source of an area's economic dynamic.

Services are firmly embedded in the urban–rural dispersal process of 'Counterurbanisation' (Champion, 1989), which is recognised as having reached most industrialised countries of the world, including Italy. Services are involved in the process principally as adjuncts to population migration, but they may also represent new 'export base' activities. There is, for instance, a US literature on the status of work in decentralised producer services offices in rural areas. The overall position in the USA was summarised as follows:

> Changing competitiveness for jobs in manufacturing and other traditional basic sectors of the economy cannot account for the greatly accelerated levels of core-periphery net outflow that have been the dominant characteristic of interstate movement during the 1970s and 1980s. Instead, an interconnected set of activities that includes government, services, trade and construction is associated with the broadscale shifts in the geographic pattern of the United States' population.
>
> (Plane, 1989: 263)

The large growth of service employment in Europe has not been spatially distributed in the same way everywhere. Illeris (1989) notes that metropolitan areas tend to have a structural advantage in the shape of a concentration of those services that have grown faster than average, for example financial and business services. However, in his review of nineteen countries' change over a variety of periods between 1970 and 1985, it became clear that in most countries decentralisation was dominant:

> The analysis also shows that this 'structural' component nowhere has been decisive for the total geographical development in service employment. Indeed, the majority of countries show a certain decentralisation of service employment in the early 1970s and 80s. The decisive factor has been a geographical shift out of the big cities, inside many sub-sectors.
>
> (Illeris, 1989: 131)

Most sub-sectors showed higher growth rates in rural areas, small towns and peripheral regions than in big cities. Increased centralisation was rare and the majority of entries show the same direction of change as the USA, the principal exceptions being the Mediterranean countries, where growth rates in big cities were on the same level as peripheral regions.

Implications for research on Europe

Throughout the 1980s, progress was achieved in service sector research by selecting disaggregated groups of activity. This represents a creditable achievement so far, allowing for the heterogeneous nature of the sector and the longstanding weakness of the published statistics. Insofar that relations with particular types of geographical areas have become better understood, the international role of metropolitan cities in producer services attracted first attention (Chapter 7). It is necessary to extend the scope of services research much further beyond producer services and metropolitan areas:

> The present understanding of the relationship between counterurbanisation and economic changes like the shift from manufacturing to services, the growing importance of small and medium-sized firms, or processes of externalisation and internalisation of services is quite limited.
>
> (Jaeger and Duerrenberger, 1991: 117)

Specific advances have been made: the strong existing literature on business services and its decentralisation (Chapter 7); studies of the geography of new firm formation in general and business studies on the nature of small firm growth (Chapter 10); and wider recognition of the value of all services in enhancing the quality of life (Wood, 1991b). There is a noticeable absence, nonetheless, of recent comprehensive studies of service sector change, although there is a reasonable and general presumption that much of its growth is sensitive to population decentralisation.

The evidence is tentative, but there do appear to be areas of Western Europe that not only demonstrate the urban–rural drift of population, but associate this with the other North American characteristics, met together, of low paid workers, service employment growth and relatively high levels of average income. They deserve priority for research under approaches that have been outlined. However, it is necessary to take a more qualitative view of women's role in the flexibilisation of employment, which I will do in the next chapter.

SUMMARY

Despite much comment on contrasts between national cultures in the ways in which 'post-industrial' conditions are expressed, there are elements of *similarity* between Western Europe and the USA. Service employment, in increasing its total share of jobs more rapidly in Western Europe, is therefore converging on the US figures, where 73.2 per cent of jobs (1993) are in services. Women have been increasing their share of total jobs on average at very similar rates in the two continents.

Major differences are apparent:

- Some EU countries have sustained a greater dependence on women's jobs over a longer period than the USA, where women's pay is much worse than men's.
- Increases in US service employment are associated with poor increases in productivity but lower unemployment than the EU. Yet the 'junk jobs' are balanced by large gains in higher paid jobs in the USA.
- Part-time employment has proved most attractive to the capitalist sector, supported by legislation, in The Netherlands, the UK, Sweden and Denmark. While tending to grow in most EU countries (but possibly to decline in the US), its incidence is irregular.
- In general the EU male trade unionist is enjoying far better conditions than his American counterpart, and the relative growth of less unionised service jobs in the EU does therefore stand as a warning.

6

FLEXIBILITY THROUGH FEMINISATION?

This chapter will explore the changing gender division of labour and its role in reshaping livelihoods in Europe. First, what of men's loss of position in the labour market and the new processes bringing women into paid work? Second, is there any lessening of women's segregation by sector, occupation or income? Finally, are there causal relationships between women's expanded role in the workforce, the growth of non-conventional jobs and the overall flexibilisation of the economy?

If there is one thing that we have learnt so far about employment change it is that women appear to be catching up on men in relative access to jobs. This was true whether we considered factories in poor countries or offices in rich, the gradual 'deindustrialisation' of the EU job structure or the growth of service jobs. The last chapter demonstrated conclusively that women's share of total recorded jobs has been steadily rising over two or more decades throughout western Europe, as in North America, Australasia and Japan, though we noticed a slowing down in Scandinavian countries as they approach a 50/50 gender division. Great Britain is forecasting for the year 2006 a convergence in 'activity rates' from which 75.4 per cent of working-age women would be at work compared with 82.1 per cent of men; in 1971 the difference lay between figures of 56.7 and 90.7 per cent (Ellison, 1994; though some of the uptake of women workers in Britain should be discounted for the large numbers of part-timers). Many now are the current affairs programmes devoted to the 'threat' that women will, through their abilities, displace men.

GENDER ROLES

The gendering of recorded employment is widely changing in a single direction. We have identified one major cause in the expansion of service activities in the employment structure of all OECD countries. However, the gender division of labour, whether for paid or unpaid work, cannot be seen purely as a mechanical product of economic change (Redclift and Mingione, 1985) or of the needs of the capitalist economy for a more or less flexible

106

workforce (Pahl, 1984). The whole balance of women's and men's employment has to be seen in the context of that large realm of life that we may term 'the home', and of 'gender relations' within it as well as in the workplace, in the state and in the prevailing culture (Walby, 1986). Gender roles are constituted by society. There are no tasks beyond the biologically given roles of childbirth and breast-feeding that require a woman rather than a man. Many myths persist, for instance that women are more 'nimble-fingered' for certain kinds of assembly or clothing work than are men, which are as readily explained by training as by hormones.

The 'starting points' in or before the Industrial Revolution have been very different. Different regions of Germany have employed different cultural practices in the gender division of labour, resulting in persistent patterns of geographical variation from the national average activity rate, even when that has risen or fallen (Sackmann and Haussermann, 1994). The same was true in Britain as between, for example, Cornwall and Wales (Massey, 1983); there has historically been a complete contrast between regions specialising in coal, steel, shipbuilding and engineering, on the one hand, and those specialising in textiles and clothing on the other (Marshall 1987), in which a great variety of systems of child-minding were developed for working mothers in Lancashire (Bagguley et al., 1990).

The point here is not to say that national or regional differences necessarily withstand pressures towards convergence during the transformation of industrialised economies, but to stress that gender roles have been construed in very different ways in different situations before, during and after industrialisation; if so, then a multitude of different processes may underlie present employment patterns, their meanings and the ways in which they are changing.

Household relations

The dominance of the conventional Western household has been eroded, to the point in fact where it is a minority feature. The 1950s maintenance of industrial capitalism by a household with a working father, married to a full-time housewife looking after their two or more children, has been reduced from many directions. Women have found it possible and attractive to take up employment after the process of child-rearing, and in turn the length of break needed for it has been reduced almost to nothing, with the growth of maternity leave and crèche and nursery facilities in most European countries. More single-parent and single-person households have emerged. At the same time the length of men's working life has reduced, with longer years of study, training and youth unemployment throughout much of Europe, while progressively earlier retirement has become the norm, hastened on by deindustrialisation.

The value of women's work at home has long been included in Denmark's estimates of national figures of GDP – but few other countries do this (Waring 1989). The British Census of Employment counts some people holding two or more (usually part-time) jobs twice over, but may exclude others, explicitly in home-working and domestic service. Yet Gregson and Lowe (1994) have stressed the rising importance of paid domestic workers, especially nannies, for the growing numbers of professional women at work in contemporary Britain. A large proportion of work in poorer countries lies outside statistical records, in the 'informal' sector. In much of Southern Europe, women's work for the market has never been fully absorbed in the formal sector, due to the continuance of unpaid work on family premises and to a wide range of temporary and casual work (Stratigaki and Vaiou, 1994). Much of this falls outside the conceptual armour of Eurostat, based on an industrialised Europe employed in full-time, unionised factory work.

CHANGES IN MEN'S RELATIVE POSITION

Masculinity established its difference by enforcing difference, by the exclusion of women. Unemployment denies that difference its institutional framework. The social space men inhabit becomes solely local and domestic, and that is the space they share with women.

(Campbell, 1993: 202)

In assessing youth violence in the 1990s, Beatrice Campbell sees the identities of certain British 'lads' thrown into confusion by unemployment and the resort to crime, which has been described as the fourth biggest industry. The effects of deindustrialisation on numbers of men's jobs are a familiar theme, but, putting it more strongly, McDowell (1991: 408) argues not only that capitalist industry no longer needs a dependable, lifelong male employee, but that, with peripheral low paid work becoming more general, 'we are all becoming "women workers" now'. This is bound to jar with systems of 'patriarchy', i.e. the dominance of men over women, which varies in time and space with regard to the precise 'gender contract' or system of relations between men and women. Walby (1986) identifies six arenas in patriarchy in Western Europe and North America over the last couple of centuries. These are to be found in (i) paid work, (ii) the household, (iii) the state, (iv) male violence, (v) sexuality, and (vi) culture. In each arena men exploit and dominate women. The sexuality and work of a housewife are seen to be appropriated by men as a basic part of the functioning and reproduction of our society. She may be subject to domestic violence and in any event is not awarded with money for this labour; and if she ventures into paid employment, *much of the subordinate role is carried over* in a form of public patriarchy, with lower pay, status and power in the workplace, possibly reinforced by further violence in the form of sexual harassment. The state and culture reinforce this public patriarchy.

If we are concerned about changed patriarchal relations then a number of strands of theory are relevant. Mincer's (1962) human capital theory of women's work suggests that up to a point women partners have a choice between working in the home and for income. They may elect to work at times of marked expansion and high incomes, but may also work more in times of depression when men may be out of work, or by extension due to the stagnation of men's incomes, as reported from the USA by Pratt and Hanson (1991). In this way we can see how restructuring may increase the supply of women's labour, and accelerate the process of convergence between the sexes in the labour market. This is opposed to the more cyclical concept of Braverman's (1974) industrial 'Reserve Army of Labour', in which a group of workers, characteristically made up of women, are called up to join the labour force when needed, and discharged equally fast when the demand has fallen away. We will see that this tends to be disproved by recent recessions, to the extent that the downswing has dispensed with men and the upswing has recruited women.

There is agreement that the dragging down of men in their relative position does not confer any benefits on women. Evidence is that gender relations in the home have changed only very slowly; even where the man is unemployed and the woman is working, the major domestic tasks still tend to be done by women. The great majority of single parents are women – 92.6 per cent in the British 1991 Census, when 17.4 per cent of children were in households with no wage earners. Polarisation between 'job-rich' and 'job-poor' households was seen in the evidence, also from 1991, that 35.7 per cent of households had two or more adults at work, 28.7 had one, but 35.6 (including pensioners) had none. Less than a quarter of households comprised two adults with dependent children. Some of this diversification is due to retirement and early retirement, and some to the growth of 'dual career households', which may represent a variety of types, some having a lead career and others not. In Europe the UK followed Denmark in having the greatest number of households favouring the egalitarian type of arrangement, where male and female partners had equally absorbing work, while in Germany, Ireland and Luxembourg the number of those preferring the 'traditional' type exceeded those preferring the egalitarian type (Kiernan, 1992).

INTERNATIONAL TRENDS FOR MEN – NO MORE 'JOBS FOR THE BOYS?'

Women already dominated in the agricultural labour forces of the old Soviet Union and several African countries in 1980. In poorer countries women have had to pick up the effects of economic crisis despite or because of the male bias of Structural Adjustment Programmes (Elson, 1991). The same has happened with deindustrialisation as when, after factory closures in

Tanzania, 'as real incomes dropped for men, women were therefore more likely to initiate income-generating activities, frequently making them the main breadwinner in the family' (Rowbotham and Mitter, 1994: 21).

This unusual report is in fact not untypical of more isolated towns facing closures in the rich world. Overall trends are more complex. Women's jobs in all countries show the greater gains, but demographic variation and cycles of growth have meant that men's employment has increased in some countries and areas in absolute terms, but principally outside Europe: the USA, Canada and Japan (OECD, 1994a). The countries that have shown an actual decline of men in the labour force over the last two decades were the UK and Belgium.

The proportion of men in the labour force (employed plus unemployed) decreased throughout the industrialised countries, from a typical 90 per cent in 1971 to 76 per cent in E15 in 1994, with the most striking reductions in France where the figure fell to 73 per cent. If we add the unemployed to the economically inactive, to produce the 'non-employed', we may demonstrate a vast increase of non-employment in rich countries, as shown in Figure 6.1. To avoid complications from the expansion of youth education and from early retirement, the figure is confined to the age groups 25–54. Even so, we can see that, for example in the UK and Spain, the proportion of men in their prime who are not employed has risen from 5 or 6 to about 18 per cent, and that non-employment has increased more uniformly than unemployment.

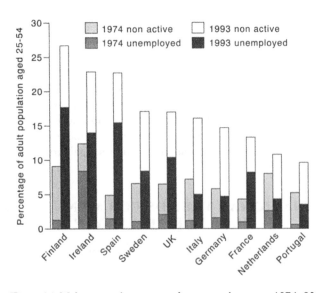

Figure 6.1 Male unemployment and non-employment, 1974–93
Source: OECD, 1994a

'National differences disguise a feature that all developed countries share, the collapse in the demand for unskilled men's labour in manufacturing industry' (Balls, 1994: 120). CEC (CEC, Eurostat, 1994) data confirm that among the unemployed men in the EU well over half had lost their last job from industry (the opposite of women, of whom two-thirds came from the service sector). Unlike the last recession of 1979–82, data on UK redundancies are now broken down by gender. Very clearly, men are twice as likely to have lost their jobs as women; at the peak of the last recession in the spring quarter of 1991, 22.9 per thousand men were declared redundant compared with 11.8 per thousand women; despite the much greater vulnerability this time of parts of the service sector, the difference is partly due to the more temporary characteristics of women's contracts. Redundancy is one of the factors that has undoubtedly led to the rise of self-employment to a level of 3.2 million (Spring 1996), three-quarters of which are men.

Unemployed male youth

If Balls (1994) can refer to high wage, low skill and 'good' men's jobs yielding a 'family wage' as becoming a thing of the past, leaving a residue of poorly educated, unskilled men in Europe and the US, and many sociologists refer to the decline of the mass-production worker, how much more missed was the opportunity of the recruitment and training for school-leavers.

The rather sketchy US evidence suggests that badly educated young men can triple their take-home pay through crime . . .

> The UK is increasingly taking the US route (to) high vacancies and rapid female employment growth alongside a rising pool of non-employed, and increasingly unemployable, men 'active' outside the system.
>
> (Balls, 1994: 126).

We saw the consequences among white unemployed youths in riots in council estates of Cardiff, Oxford and Tyneside in 1991 and Bristol, Coventry, Salford, Blackburn, Burnley, Huddersfield and Carlisle in 1992. In a magnificent report on this crisis for masculinity, we find the following:

> 'These men live in a twilight world,' said one of the men's lawyers. 'They're lying around on the sofa in their boxer shorts, watching videos; they have tea when it's put in front of them; they go out TWOC-ing and burgling. When the men get in trouble, or when their wives want them out, it is their wives and mothers who make the arrangements. . . . What is absolutely astonishing about these tough men is that they have to have their slippers under some woman's bed. The men cannot make out on their own. The reality is that children in this

community do not grow up seeing men do any of the coping, caring or standing on their own two feet.'

(Campbell, 1993: 178)

(Note: TWOCing is joyriding in cars Taken Without the Owner's Consent)

It is clear that many younger men are trapped in a vicious circle of moving between schemes and unemployment; at any one time there are about 300,000 people on government employment and training programmes, about two-thirds of which are men. New statistical series (*Labour Market Trends*, 8.3–8.6, monthly) show clearly that the longer the period of unemployment before training, the less the chance of gaining a job thereafter; these figures are worse than average for youth, better for women.

Unemployed older men

There is, however, a general problem of unskilled men of all ages. With the growing belief that their unemployment is no longer cyclical, there is less real point in keeping them in touch with the world of work through temporary schemes. They themselves meanwhile are reluctant to take 'women's jobs' because of pride, or because pay rates are too low to justify coming off social security. Many low skilled workers have left the labour market altogether by signing up for long-term sickness benefit.

A major factor in explaining the fall in men's activity rates during the early 1980s was the high levels of unemployment which discouraged people from entering or remaining in the labour market. Recent European Labour Force Survey data show marked reductions in the proportions of older men at work after this last recession. Governments have tried to counter 'ageism', a prejudice among employers against recruiting or retaining older workers, but contradictory attitudes are held both in firms and in government itself. Recent indications are that older men are taking advantage of pension opportunities, especially when state and occupational pensions are available.

Until recently we knew relatively little about people of working age who were not in the labour force: in the statistical terms of Table 6.1, those who wanted a job but who had not sought work in the previous four weeks. In all, there were 2.2 million people in this category in Great Britain, nearly two-thirds of them women, half of them looking after domestic affairs. There were also more than half a million who were counted as long-term sick and disabled, 350,000 of these being men.

The geography of these groups has previously been neglected but adds considerably to our understanding of regional poverty and change. Activity rates for men in the 1991 Census varied between 79.0 to 80.1 per cent in the counties of Berkshire, Buckinghamshire and Bedfordshire, on the one hand, and 64.8 to 68.0 per cent in Dyfed, Mid- and West Glamorgan on the other. A 'limiting long-term illness' was declared by no less than 20.4 per cent of all

Table 6.1 GB economically inactive, including 'discouraged' workers, by reason (thousands), winter 1995/6

	Male	*Female*	*Total*
Total economically inactive	6,055	10,601	16,656
Does not want job	5,133	9,186	14,320
Wants job but not seeking	842	1,315	2,157
Discouraged workers	75	54	129
Long term sick/disabled	350	212	562
Looking after family/home	71	676	747
Student	143	129	272
Other	204	243	447
Seeking but not available to start	100	79	179

Source: *Labour Force Survey, Quarterly Bulletin*, 16, June 1996

men in Mid-Glamorgan, 19.0 in West Glamorgan and 17.4 in Co. Durham. These men are excluded from unemployment rates, and it is only when we consult data for GDP per capita that we realise that, for instance, Co. Durham is among the most deprived counties of England.

We can see, therefore, how the legacy of male-based industrialism is providing a burden both to older men and to boy school-leavers in many parts of the country. However, Table 6.1 nonetheless showed that 9.2 million women compared with 5.1 million men were outside the labour market, and we can now return to the central issue of supply and demand for their labour.

THE PROCESSES THAT ATTRACT WOMEN TO PAID WORK

'What is new is the large-scale employment of married women from all social classes for most of their married lives on a full-time basis' (Pahl, 1984: 85). Clearly, the erosion of incomes and opportunities for men is only part of this story of change. Many see the rise of contraception and labour-saving domestic appliances as key elements, but I find the close correlation, found in Chapter 5, between the rise of women's jobs and that of service employment at least as compelling; but it is necessary to look at both supply and demand for women's paid labour.

Pahl (1984) was prominent among those who stressed that the later Victorian and Edwardian confinement of women to the home represented the aberration, and that present directions of employment change are therefore restoring some previous variety to women's lives. These directions, in France for instance, originated before the full influence of contraception and the domestic-appliance industry were felt. In Scandinavia, the key feature was that of political and cultural support for equal opportunities through pervasive public child care and formal regulation of wages, although women still have a pattern of lower value jobs even when they have nearly half of them (Forsberg, 1994). In Eastern Europe, the ideological commitments of governments to equal economic and social rights and the high demand for labour made the encouragement of women's work a central feature, such that women's participation was extremely high by world standards, involving 70 to 80 per cent of women of productive age (Bodrova and Anker, 1985). However, the pattern was inherently inequitable, and, in East Germany for example, has failed to survive the end of Communism (Quack and Maier, 1994). Countries seen to have lower dependence on women's work in Chapter 5 (Figure 5.4) include Greece, Italy, Spain and Eire, which have experienced Catholic Church domination. Stratigaki and Vaiou (1994) argue that, in connection with the late industrialisation of Southern Europe, much of women's productive work lies outside the realm of wage labour in unpaid family work etc. The immediate explanation for Eire's position lies in a legal bar on the employment of married women until 1973. The field is then a rich one for the interplay of social, cultural, religious, institutional and political features when we compare the respective levels of women's work in different 'rich' countries at any given date.

The statistical material of Chapter 5 was silent if we wanted to explain the inherited pattern at any one date. However, it was when we came to the issue of change that we found parallels between different kinds of country. Growth in women's jobs might be termed 'structural', because they are achieved not principally by women taking over men's jobs but simply by the expansion of sectors or occupations in which women were already well represented. Christopherson (1989: 140) saw how in the USA 'with growing demand in service jobs, young married women went into the workforce in large numbers . . . between 1970 and 1985'. In the UK from 1954 to 1991 most of the increase could be 'linked to changes in industrial structure tending to favour those industries . . . in which employment is predominantly female (the "industry" effect)'.

Compared with this the 'gender effect', involving women capturing a rising share of jobs in other sectors, was relatively modest, though notable in financial and business services (Wilson, 1994). In a study of the City of London, McDowell and Court (1994a) found that women had a low proportion of jobs and for most of the 1980s did not gain on men (but see Chapter 7 on 'back offices' elsewhere). This will appear disappointing to liberal feminist campaigners and women student readers, but we have to accept that many theorists see demand as dominant, the French 'regulationists' for example and those who fall back on the concept of an Industrial Reserve Army. Another dominant factor is that a large number of lower paid women workers are performing literally a personal service of the kind that cannot be automated but which is 'traditionally associated with the socially constructed attributes of femininity' (McDowell and Court, 1994b; 1399), and which unemployed men would be unlikely to want to do under present constructions of gender.

Interaction with the rest of the household

It is at this point, however, that we are reminded of interaction in the household. Not only may changed domestic arrangements have facilitated women's entry to work, but so too may *men's straitened circumstances* have prompted it. Thus, in areas suffering deindustrialisation, the growth of women's work in services has played a prominent role in effecting the 'restructuring thesis', as shown in Lancaster by Bagguley *et al.* (1990). In the steel closure town of Consett, Co. Durham, Hudson *et al.* (1992: 51) found 'whether in homes for the elderly, pubs or supermarkets, job growth has predominantly been composed of low skill, low wage, part-time or casual jobs, overwhelmingly taken by women'. Hanson and Pratt (1995) argue, given the massive increase of women's jobs in the US, that the need is to understand the new arrangements adopted by 'family-households', including residential location.

Many women are engaged in *caring roles*, paid and unpaid, both inside and outside marriage. For example, 53,900 women staff were recorded by the GB Census of Population, 1991 (Table L03) as (living-in) residents of Health Service hospitals and homes, hotels, educational establishments and private residential homes. Many more women are committed on virtually a full-time basis to care in the home of the elderly, and of adults and children with disabilities and mental handicaps:

> The evidence, then, is unequivocal. While the family, where it exists, still cares for its elderly members within the family, it is wives, daughters, daughters-in-law and other female relatives who shoulder the main burden of responsibility. Moreover, when the carer has taken on the burden she is likely to receive little practical support from other relatives.
>
> (Parker, 1988: 504)

This factor, especially when combined with the erosion of the welfare state (McDowell, 1991), may be seen as the principal constraint on the growth of women's employment.

Many other factors in the household and demographic change are conducive to women taking jobs. The number of children per woman, projected over the *total fertility* period, has fallen in E12 from 2.40 in 1970 to 1.48 in 1992 (2.43 to 1.80 in the UK). While divorce in the UK more than quadrupled over twenty years, France and Germany both experienced a threefold increase in divorce rates, and related increases in the number of working-age single-person households.

Economic activity rates

Recorded rates of economic activity increased in all the countries of E15 in the years 1975 to 1994, due to the increase of women's activity rates (and rates of actual employment), both moving in the opposite direction to the average for men (Meulders *et al.,* 1993). By 1994, no less than 78.2 per cent of women were working in services, and 30.3 per cent part-time. In the UK, employment rates rose more rapidly for women with children than for other women. Nevertheless, this only served to place it more firmly in the middle rank of EU countries for mothers' employment rates (Harrop and Moss, 1994).

Figure 6.2 provides age-specific activity rates for both sexes in all twelve countries, and shows how women (shown by dashed lines) are returning to work sooner and sooner after the births of children, leaving less of a dip in average age-profiles (as in Germany and the UK at age 30) and eventually no dip at all (as in Denmark). In most EU countries the age of the youngest child now makes little difference to mothers' employment rates, except in Germany and the UK. Lone mothers are more likely to be in work than other mothers in most countries. The UK was one of three exceptions, with a steep rise from a low activity rate being evident in the three age groups shown in Table 6.2; given the poor provision of child care, these women

Table 6.2 GB employment rates by family type and age of youngest dependent child, people of working age, winter 1993/4

All	*Thousands in employment*	*% of population*
All men (16–64)	17,814	75
All women (16–59)	16,344	65
All mothers (16–59)	6,522	58
Mothers living in couples		
youngest dependent child 0–15	5,399	62
0–4	2,601	51
5–10	1,607	70
11–15	1,190	76
Female lone parents		
youngest dependent child 0–15	1,123	39
0–4	543	25
5–10	359	45
11–15	221	63
Male lone parents		
youngest dependent child 0–15	140	54
0–4	34	53
5–10	52	49
11–15	54	59

Source: *Labour Force Survey*, reported in Harrop and Moss, 1994

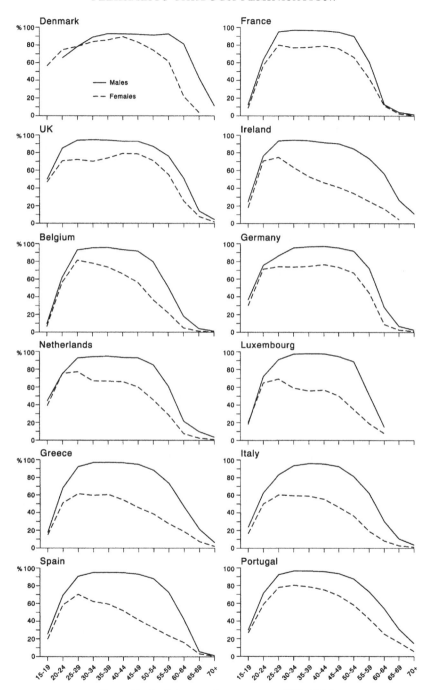

Figure 6.2 Activity rates by age group, E12, 1994
Source: CEC, Eurostat, 1996c

117

have great difficulty in sustaining anything like their previous patterns of employment. Among women as a whole the biggest rise in labour market participation has been among those with children under 5, up from 37 per cent in the spring of 1984 to 52 per cent in the winter of 1993/4 (Table 6.2). Participation rates of mothers, unexpectedly, increase with the level of the partner's pay. As the *Financial Times,* (7/8 January 1995) recommended an income of £40,000 p.a. as a minimum for taking on a nanny, it is not surprising that surveys variously show only 2–4 per cent of couples using a live-in or other nanny. Thirty per cent rely for child care on the partner, 25 per cent on parents (in-law), 15 on other relatives and friends, 11 per cent on a childminder and 8 on a nursery, nursery school, crèche or playgroup (Marsh and Mackay, 1993).

New kinds of household require new concepts and accounting techniques, which may have to cope with different labour market behaviour from the offspring of these households (Elias and Hogarth, 1994). Clearly we have discovered continuing substantial barriers to women's employment in different places and kinds of household. However, they are best seen as offering different constraints on the dominant trend, which is for the service sector to want more women workers. It is true that those women who have returned to the labour market after extensive breaks have been disappointed by their level of employment, connected with the part-time status so useful to employers' needs for flexibility (Martin and Roberts, 1984).

Summary

We agree then with Walby (1986) and others that women's increasing employment is a function of shifts in the nature of employment towards the sectors that have traditionally relied on low paid women workers – the structural argument – rather than of increasing supply being the dominant factor. 'Changes in the industrial structure of employment in favour of services will tend to provide job opportunities which traditionally have tended to be regarded as typically female' (Lindley, 1994: 27), but men's loss of jobs or income may speed up women's take-up of jobs, and 'while the dominance of demand-side influences must be recognised, the potential impact of women's behavioral and attitudinal changes on employment trends must not be minimised' (Rubery and Fagan, 1994: 29).

DIRECTIONS OF WOMEN'S EMPLOYMENT CHANGE

There was recently a woman pilot at work on the Newcastle-upon-Tyne to Paris flights, and it is commonplace now to spot women workers in the male bastion of British railways. Our purpose in this section is to see how the processes we have described are expressed in the human geography of 'richer' countries: the gendering of industrial sectors, occupations and pay in urban and rural areas.

Gender segregation by industry

'Once the sex-ratios in the occupations of particular industries have been set they are usually quite resistant to change' (Walby, 1986: 88). As the structural swing towards services is largely responsible for women's increased employment, then any further change in the sex-ratio within individual branches of services is a 'bonus'. Table 6.3 shows how the EU and the USA reached such levels through increases in women's share of employment in services but not other sectors. In E12 women have provided just above a third of the agricultural workforce since 1965, and also 23 per cent of industrial workers.

Of course this shift itself may in turn result from differential changes in the service sector; i.e. 'male' sectors like transport may be economising on staff while more 'female' employing sectors such as banking and finance are expanding. Since 1965 all five main woman-employing sectors in the US and UK showed a steady swing to greater women's employment, led in 1991 by community, social and personal services (60.6 per cent women in the US, 62.8 in the UK) and followed in the US by finance, insurance, real estate and business services, and in the UK by distribution, restaurants and hotels. In the UK manufacturing sector the recruitment of women to the labour-intensive branch factories of TNCs in the 1970s (Massey, 1984) was offset by the disproportionate rundown of women's jobs in food, textiles, footwear and clothing, electronic assembly and instrument engineering. Otherwise only hotels and catering significantly reduced their proportion of women workers, although distribution, personal services and business services appear to have reached a peak in their proportion of women.

Table 6.3 Women's share of total employment by sector, 1975–94 (per cent)

		Agriculture	Industry	Services	Total
USA,	1975	17.1	23.3	48.7	39.3
	1985	20.1	25.8	52.6	44.1
	1992	20.9	26.2	53.4	45.7
E12,	1975	34.5	23.2	42.6	34.2
	1985	34.6	23.2	47.5	38.2
	1994	34.4	23.2	50.1	40.8
UK,	1975	20.3	26.2	50.0	39.0
	1985	19.3	24.0	51.9	42.3
	1995	23.0	24.0	57.5	49.4

Source: CEC, Eurostat, 1995b; OECD, 1994a; *Labour Market Trends*

Gender segregation by occupation

Crudely, women are segregated into lower paid service occupations as well as into the service sector. More subtly, women suffer patriarchal discrimination in many forms.

Much of the spatial division of labour (Massey, 1984) could be better understood if occupational data were available more often. The gendering of work is fairly similar in all member states, except among the self-employed, and in the UK where the general position was slightly more favourable to women. Significantly more men than women were classified as 'managers' in 1992. For 'professionals' the proportions of men and of women were broadly the same. In other activities the differences were much more pronounced, with much higher proportions of women than men employed as clerks and service workers.

Structural changes are expected to continue to favour highly skilled workers irrespective of growth rate and recessions. The past record (Figure 6.3) shows how the growth of professional and technical jobs for women was the most persistent. At the same time, employment of women clerical workers went into decline in 1990–1.

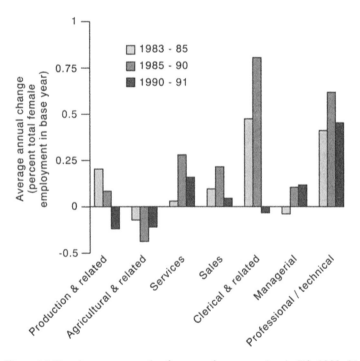

Figure 6.3 Employment growth of women by occupation in E9, 1983–91
Source: CEC, 1994a

'In the 1980s there is no evidence to suggest that any real changes have occurred in the degree of segregation' in the EU (Perrons, 1994: 1210, using an index of dissimilarity). Occupational segregation declined slightly in all of the EU countries, with the exception of Ireland, and differences between countries suggest that policy may have had some effect. Even in Sweden, a most advanced entrant to the EU, the very high activity rates nonetheless leave women doing basically an inferior set of jobs to men (Forsberg, 1994). In the case of the UK, segregation has been reduced a little by women's increased share of work in the majority of occupational groups, especially managerial. Nevertheless, McDowell and Court (1994b) stress how gender divisions of labour persist in the City of London. Rubery and Fagan's (1994) use of an index of occupational segregation for Britain shows only minor fluctuations in the 1980s.

Gender and pay

It is a remarkable achievement of capitalist patriarchy to pay women less than men even for the same work, often achieved by designating women's sections of a firm as different 'occupations' and therefore pay groups from men. 'Most fundamentally, however, the persistence of horizontal and vertical employment segregation means that women are doing different jobs from men, and traditionally male skills are still more highly rewarded than female ones' (Perrons, 1994: 1209). As mentioned in Chapter 5, there are very few Eurostat data available on the fast-growing service sector, so comparison has to be restricted to fairly old evidence for manufacturing, such as are shown in Figure 6.4.

Following EC directives of 1975, there is a general tendency for women's earnings to catch up slowly on men's, so long as overtime is not taken into account. However, the issue of unequal pay has been devolved through national decentralisation of wage determination to individual employers, many of them small in the case of women (Perrons, 1994: 1209). The figure shows a setback for women's earnings in the 1980s in Denmark and the UK. From 1988 to 1994, however, a recovery occurred in the UK which must reflect the recession experience of men, in that the ratio of women's to men's full-time weekly earnings (in all sectors) increased from 66.8 to 72.2. There was remarkably little geographical variation in the ratio.

A changing balance of urban and rural

One of the most geographical dimensions of change comprises the urban–rural. How important are urban–rural contrasts? There is debate, for instance in Germany, whether the understanding of spatial variation in activity rates should be focused on urban–rural or regional differences. In truth, many spatial differences in the geography of gender are disappearing under

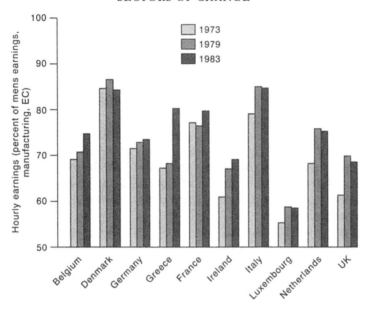

Figure 6.4 Hourly earnings of women as a percentage of men's earnings,
manufacturing, EU
Source: CEC, Eurostat, 1991

widespread convergent changes. Gregson and Lowe (1994) were disappointed to find that professional couples were behaving much the same in Tyneside and the M4 corridor.

Sackmann and Haussermann (1994) argue that the south–north divide in Germany, with high women's activity rates in the south (Figure 6.5), is a longstanding feature, predating the post-war economic growth of the south. So they conclude that 'regional cultures', especially constructions of gender expressed in occupational segregation, have a much deeper effect than just the geography of the service sector. While unemployment of women is generally much lower in the south, rural sub-regions show the lowest unemployment rates and the highest shares of employment for women, but only the second highest levels of activity rates.

Population dispersal to rural areas has been identified in most countries (the 'counterurbanisation' of Champion, 1989), and Townsend (1986) has shown how the dispersal of population to rural areas has been the main identifiable cause of the growth of part-time employment, much of it of women in services. This leads to the main present-day distinction between urban and rural in the UK. This used to be one of low opportunities and pay for women in rural areas, contrasting with the strongest growth in conurbations; now the main distinction lies between a strong dependence on full-time women's jobs in London, and part-time jobs in rural areas. Table 6.4

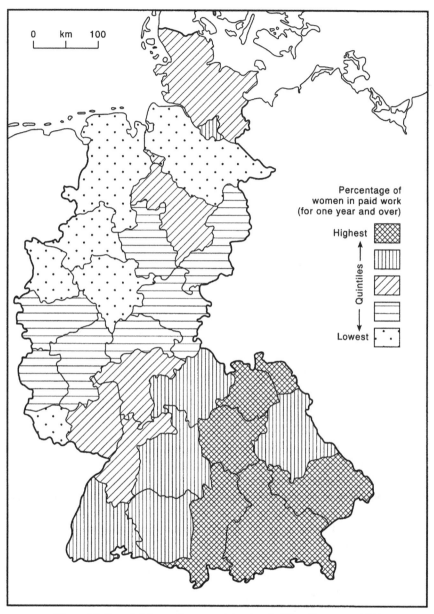

Figure 6.5 Female labour force participation in West Germany, 1991
Source: Sackmann and Haussermann, 1994

shows how there is a systematic gradation in these types of activity across the range of sub-regions, with very little variation left around women's crude total activity rate of 67.6 per cent.

Table 6.4 GB female economic activity rates, aged 16–59, by urban–rural types of district, 1991 (per cent)

	Full-time	*Part-time*	*Self-employed*	*Unemployed*	*Total*
Inner London	42.1	10.9	4.6	8.7	67.3
Outer London	42.1	17.2	3.6	5.0	69.0
Principal cities	34.2	19.9	2.4	7.0	64.7
Other metropolitan	34.7	22.5	3.0	5.3	66.4
Large cities	36.8	22.0	2.8	5.3	67.7
Other cities	36.5	23.5	3.5	4.5	68.9
Industrial districts	34.8	23.1	3.3	4.5	66.7
Districts with New Towns	38.2	22.0	2.9	4.6	68.6
Resort, port and retirement districts	33.4	23.9	5.8	4.1	67.8
Mixed urban–rural	37.4	24.0	4.6	3.2	69.8
Remoter mainly rural districts	31.7	24.3	6.3	3.6	66.8
Great Britain	36.2	21.9	4.0	4.7	67.6

Note: Total includes 0.8 per cent of the female adult population (GB) who were engaged on government training schemes.
Source: Townsend, 1994, from data supplied by Champion and Dorling, December, 1992

Overall directions of change for the future

The direction of change is expected to continue. Women's work as conventionally defined through activity rates is expected to dominate change in the EU (CEC, 1995b) but the quality of jobs is expected to improve only very slowly, not helped much by the institutions of Europe.

But will recent growth sectors be as important in the next ten years as the last? Notably, finance and business services were not expected to resume a leading position in overall job growth in 1990s Britain, but to yield it to 'miscellaneous services', activities such as hairdressing, and health and education (Lindley and Wilson, 1993). Recent annual changes show most manufacturing sectors, insurance and parts of distribution losing women employees. These were somewhat different sub-sectors from those expected to be affected by the UK's involvement in the Single European Market from January 1993. Lindley (1992) expected pressure in textiles, but he thought that there might be some potential for higher quality jobs in retailing and tourism.

Lindley (1994) did not think that economic arguments for the further integration of women would be strengthened by the development of the social dimension of the Single Market or the arrival of the European Monetary Union. Perhaps the best hope for women lay in the accession of rich countries amenable to a culture of women's work, notably Sweden and Finland. He felt, however, that equal opportunities programmes were still leaving women with a burden of disadvantage, especially where a career break led to part-time employment, and consequent downward mobility.

NON-CONVENTIONAL EMPLOYMENT

It is a convention only of north-west Europe that 'work' should be of a '9 to 5' weekday variety and for an employer. As we showed in Chapter 2, poorer countries provided much less clear distinctions between paid and unpaid work, indeed the growth of the 'informal sector' was mentioned because it was at times seen as the answer to under-employment. Now that unemployment and non-employment are permanent features of the European scene, we can also see much work being diluted across a range of non-conventional outlets, probably more a result than a cure to the underlying economic problem. Flexibility is shown by the different ratios of full-timers, part-timers and the self-employed in rural Britain. Beyond such statistics of less conventional employment, however, lies the whole world of the 'informal sector' (Redclift and Mingione, 1985; Pahl, 1984) or *Atypical employment in the EC* (Meulders *et al.*, 1994).

Statistically visible and non-conventional jobs

We shall now discuss non-conventional work of both sexes, before concluding with an answer to our question of whether flexibility in the labour market is mainly provided by women – and at what expense to them?

Productive and reproductive work are blurred under the international heading of *'unpaid family labour'*. This is inconsistently used between countries, but Japan reports 4.9 million people (13.0 per cent of workers) under this heading, including many women on farms who are declining in numbers, as in Southern Europe. Nonetheless, CEC, Eurostat (1996c) can still record 3.3 million family workers in E12 in 1994, 70.3 per cent of which are women.

Among the *self-employed and employers* the decline in the number of farming proprietors is more than offset by the growth of businesses in industry and (especially) services. In France, West Germany and the UK, from the later 1970s there was a striking and consistent relationship between the rate of change in employment and size of manufacturing plant: the smaller the pattern of establishments, the greater the rate of employment gain (Keeble and Wever, 1986; Storey and Johnson, 1987). Over the period 1985 to 1994, however, the self-employed showed no overall change in their share of E15 jobs overall (15 per cent), or of women's jobs (9.5 per cent). Thus the sector has expanded only in proportion to jobs at large, and women have made no greater inroads than in the labour market as a whole. They represent only 26.0 per cent of the sector in E15, and are therefore not much identified with this area of change.

By contrast, there is little gender distinction among *temporary workers*; they have expanded only very gradually to occupy a share of 11.0 per cent of EU workers in 1994 (12.1 for women, 10.1 for men).

Part-time work therefore remains the most gendered of the flexible kinds of employment. In the British Census of Employment this is defined as less than thirty hours work per week, but in the European and British Labour Force Surveys it is defined by the respondent, and averaged 19.0 hours per week in 1991 compared with 40.2 for full-timers. Pfau-Effinger (1994) argues that national variations in part-time/full-time ratios derive from women's position, especially in the family, even though it is the expansion of the service sector that is the precondition for the expansion of part-time jobs. It is clearly the numerical flexibility provided to employers that is important in serving customers in shops, hotels, catering, hairdressing and cleaning, as part-time workers are easier to adjust to customer demand. In all, only 4.8 per cent of men were recorded as part-timers by CEC (1995b), compared with 30.3 per cent of women in E15 in 1984. Thus women represented 81.8 per cent of part-timers in the EU. Lastly, women represented 58.1 per cent of the 1.3 million GB workers with *second jobs*, as recorded in the *Labour Force Survey Quarterly Bulletin*, June 1996.

Taking these quantifiable forms of non-conventional employment together, it is not surprising to find that women represent 60 per cent of the 'flexible workforce'. In turn that represents about 40 per cent now of the overall labour market among both sexes (38.5 in Table 1.1). Watson (1994) argued that in 1993 some 9.7 million people, 38 per cent of all UK workers, constituted a 'flexible workforce' of part-time, self-employed, government training scheme or unpaid family workers – an increase of 1.25 million since 1986. The percentage of all working men who are in this 'flexible workforce' had risen from 18 in 1981 to 27 in 1993, whereas among women it had remained stable at about 50. On a similar basis Christopherson (1989) esti-mated that approximately 25 per cent of all US jobs were 'flexible'.

Statistically invisible and non-conventional jobs

By its very nature, a further proportion of work is unknown to statisticians of the *Labour Force* and other surveys, in particular just because those involved may be evading social security and VAT law. Many believe that work of the *informal sector* is concentrated in deprived populations and localities, such as, for instance, Southern Italy. 'Unrecorded employment is more likely in countries like Italy, with a flourishing sector of family capitalism, and perhaps less likely in Britain, where the concentration of capital has pro-ceeded further than in any other Western capitalist state' (Pahl, 1984: 116). Garcia-Ramon and Baylina (1994), studying factory outworking in Spain, see all sorts of savings for firms, not just in premises, low wages and social security payments but also in exploitative control of women workers. In fact, the CEC (1991b) concluded that in most northern parts of the Community, the informal sector probably represented an addition of about 5 per cent or less to the level of recorded work, but somewhat more in France and Belgium, possibly reaching 10 to 20 per cent in Southern Europe.

A review by Williams and Windebank (1994) found that variations did indeed exist within countries both in the size and character of the informal sector. However, they found that across the EU affluent localities engage in *more* informal work, much of it, as was found in poorer countries, closely linked to the formal sector:

> The result is that informal work is not a substitute for formal employment. It cannot be used as a welfare mechanism, nor as a survival strategy for deprived populations and areas. Instead, the informal sector mirrors the spatial inequalities present in paid employment.
>
> (Williams and Windebank, 1994: 821)

Informal, low-paid, industrial homeworking may have more in common with poorly paid industrial work than it does with a plumber mending a neighbour's tap for cash.

Most estimates of the amount of undeclared work in Britain put it between 5 and 10 per cent of the total economy, but it is less clear whether this would represent the addition of the same proportion of jobs. Certainly the contracting out of work from major industries, even the steel industry, is carried out on a fairly informal basis which may merge into the 'underground economy'. Casual work or paid domestic labour by women is also evidenced from a steel closure town:

> In an area where jobs for women are mostly in the service sector, part-time and badly paid, it is not surprising to find women also resorting to informal economic activity. Personal service jobs 'on the side' are not only taken by women whose husbands are unemployed – although this is probably the most desperate of situations. People employ women to do their household cleaning for £2 an hour or less because they know 'she needs the work, we can get her cheaper than we can get a home help in'. In view of the low wages within the formal part of the service sector, it often seems more rational to register unemployed and work part of the day informally and illegally.
>
> (Hudson *et al.*, 1992: 53)

However, Gregson and Lowe (1994) see the revival of domestic labour for the prosperous middle classes as a secular trend. This employment of cleaners and nannies is not yet widespread – they say it is specific to about a third of dual career partnerships. This is to be distinguished from another solution open to carers, and others of both sexes, that of *home-working*. Hakim (1988) reports that home-working is spread wide across nearly all sectors and employs a wide cross-section of people, with some bias toward graduates of higher education but also toward poor pay and conditions.

In total, 1.7 million (in England and Wales, 1981) or 7 per cent of the workforce could be defined as home-based workers, but the majority of them were people living at a workplace or working from home, like certain lorry drivers. Of the total, only 4 per cent were recorded as child minders, and 9 in manufacturing. There has since 1981 been general interest in the growth of 'teleworking' (Huws, 1993) because the combination of computer and telecommunications technologies has made it technically possible for large numbers of workers whose jobs involve information processing to work remotely, at terminals that could be in their homes. It was being confidently forecast that IT would transform patterns of work, freeing people from the office and from problems of child care and enabling them to reduce travel congestion by working at home.

This is certainly a live direction of change for us to consider, but it is not in fact a strong trend, or necessarily a very attractive one. Freelance writers and publishers' staff certainly work from home and are sometimes free to move to an 'outer rural' area. However, Huws (1993) reports that, although one employer in ten employs some form of home-based worker, and one in twenty at least one teleworker, these make up only 1 per cent of the workforce, of whom less than one worker in 200 is a genuine teleworker. There are significant concentrations in South-East England and financial and business services. A minority of schemes are characterised by problems of isolation, insecurity and low pay.

Flexibilisation – a summary

Perhaps not surprisingly, the consideration of statistically less visible types of work has not added much to our total of non-conventional jobs. Telecommuting is as yet a minor subset of home-working, and most of that in turn overlaps with and is recorded already under self-employment, part-time or temporary work. The flexible economy certainly can be recognised in conventional statistics.

It is not, however, recognisable in the precise terms of the literature reported in Chapter 1, where male geographers demonstrate a curiously out-of-date concern with manufacturing. The numerical flexibility represented by part-time and self-employed work represents smaller proportions of the industrial sector, merely 16.9 per cent in the EU in 1994, and 21.4 per cent in the UK. The flexible firm literature would also suggest that the use of temporary, part-time and self-employed labour would reflect a deliberate managerial strategy to create a flexible workforce, rather than this being a result of high unemployment and reduced trade union influence. A study in the UK (McGregor and Sproull, 1992) showed that employers themselves did not recognise a conscious 'flexible firm' model. The great majority took on non-conventional staff for the traditional reasons of dealing with tasks with limited time inputs, matching staffing to demand and of taking advantage of the preference of some kinds of worker for part-time work.

CONCLUSION – WOMEN'S RELATIVE CONTRIBUTION TO THE FLEXIBLE ECONOMY

Flexibilisation can be seen as central to the massive recruitment of women across the USA and Europe; 'most women in the United States are now lifetime wage earners albeit frequently in "flexible jobs"' (Christopherson, 1989: 40). It contributes to women having a 50 per cent greater rate of job turnover than men in the EU today (CEC, 1994a). Although firms' needs for flexibility come first, the availability of women has clearly contributed to the volume of part-time work, in which they normally make up over 80 per cent of employees.

In Britain it can be argued that the whole of women's increased participation rates may be attributed to part-time activity. Post-war growth only restored the full-time activity rates of the peak Victorian years of 1851 to 1871; without the growth of part-time work as from the 1980s, the growth of women's employment would indeed be something of a myth in Britain (Hakim, 1993). It is tempting to view part-time growth as a straitjacket imposed on women, and to dismiss endless survey reports in the official *Labour Market Trends* that say that three-quarters or so of women workers prefer part-time to full-time work. However, this view is supported by Pinch and Storey (1992) who in a survey of Southampton found considerable satisfaction among part-time workers, at least given the alternatives available, and that their part-time work could not be universally designated as peripheral.

Women's position is subordinate among employers and the self-employed (Campbell and Daly, 1992). It would be far more correct to say that the flexible economy is associated with women's part-time paid work in services rather than with male manufacturing.

Low-paid women's work as part of the polarisation of society?

Even the most cautious text will identify the problem of low pay for women and question whether EU Social Policy can provide a remedy. The issue is accentuated by the increasing proportion of single women and mothers and the question of whether the incidence of low income among households is leading toward the 'polarisation' of societies. There are opposing views on the question of polarisation. For instance, Stanworth (1993) argues that peripheral, casualised labour will continue to be a growing feature of the labour market. Until now this sector has been a dumping ground for cheap labour, especially for women who are willing to trade pay for flexibility, particularly during years of childbearing; the drawing of men into the casual-ised workforce only adds to the political challenge of this problem. In a sustained thesis, Sassen (1991) argues that the leading 'global cities' have, through abandoning manufacturing and expanding in financial and business services, developed a polarised workforce with growth of jobs at the top and bottom of the income scale, a decline in the middle and the creation of large numbers of low income jobs. This relates to Hutton's (1995) 40/30/30 society.

Hamnett (1994), however, thinks that even if Sassen is right about New York and Los Angeles (and there are fresh reasons for doubt on that score), the situation there is altered by the availability of immigrants. In The Nether-lands and Britain, the statistical evidence is that the expansion of employ-ment has been in higher skilled jobs, leaving the numbers of unskilled jobs much where they were. For instance, in The Netherlands the increase of women's and men's jobs was about equal in numerical terms, but pro-portionately stronger for women; 'women are by no means being concen-trated in low-skilled jobs in the four city regions' (Hamnett, 1994: 414). In Britain, extensive work by Webb (1994) attempts to fill the gap in most published statistics for income distribution by household. The over-representation of women at the lowest income levels proves less marked when account is taken of the incomes of other household members, but two-thirds of the adults in the poorest households are women. This still represents a considerable improvement on the situation of twenty years ago, when one pound in every four that came into UK private households was earned by a woman. In 1991, that proportion had risen to one pound in three, due to women's greater participation.

There is still a long way to go. Ten per cent of adult full-time women in the UK earned less than £139 per week, and 23 per cent less than £170 in 1994 (New Earnings Survey). Until 1993, 2 million women workers in Britain had their minimum rates of pay set by wages councils which were then abolished, leaving the country as the only member of the EU with no minimum wage.

The 1990s have been dubbed 'the decade for women in Europe', and the Social Charter recognises that part-time workers should enjoy the same pro-tection and employment rights as their full-time equivalents. The UK was, of course, the only country to 'opt out' of the Charter, but was forced to accede the same redundancy and dismissal rights as full-timers in 1994. Despite setbacks, then, the position of women on average is slowly improving. There are even areas of Britain that record more women employees than men - not only Essex, Sussex, Devon, Cornwall and Hereford and Worcester, but also Merseyside, the Scottish Borders and Lothian (before counting the self-employed and discounting part-time hours). On the other hand the eco-nomic inactivity of 62 per cent of lone parents and men's loss of jobs means that 17.4 per cent of children (as of the 1991 Census) were in households with no earners. This proportion was at its greatest in the northern conurbations.

> The contrast between the households with money being productive and busy and the households without money being unproductive is the overwhelming conclusion of the empirical part of this book.
>
> (Pahl, 1994: 336)

SUMMARY

- Women's paid work must be seen in the context of changes in households.
- The severe effects of unemployment on men must be analysed. The proportion of men who could be considered part of the 'flexible work-force' of the E15 had rose from 18 per cent in 1981 to 27 per cent in 1994.
- However, by 1991 women performed over 60 per cent of the non-conventional jobs in both the EU and the UK.
- The flexible economy is associated with women's part-time paid employee work in services rather than with men's work in manufacturing.
- Women's share of jobs and income in very different occupations and industries is advancing only very slowly.

7

'PRODUCER SERVICES' – IN THE LEAD FOR GROWTH?

I now turn to some of the fastest growing and better paid services, before proceeding in the next two chapters to dynamic change in two more labour-intensive sectors. In the world of jobs, nothing can rival the dynamic growth and social impact of the swing to 'producer services'. I begin the chapter with definitions before considering how the sector may concentrate jobs in particular regions and locations through providing services for other areas. There are tendencies both for concentration in metropolitan areas and for decentralisation from them, which raise unexplored questions about the quality of jobs in 'back offices'. Although computer services provide the example of a notably buoyant sector, there are also questions as to whether the early 1990s recession has not signalled the end of growth in large parts of financial services.

The growth of producer services was the centrepiece in the restructuring of North American cities. These services have now attained global significance in locations such as Frankfurt, Tokyo and now even Bombay, and lay at the heart of a surge of income and wealth in London of the 1980s, with most of the higher paid jobs still being for men. They can also be seen as prime examples of 'flexible specialisation' (Wood, 1991a) because other 'producers' substitute purchase of these services for the task of performing the same functions in-house.

DEFINITIONS AND GROWTH

Being the opposite of 'consumer services', these 'producer services' are sold not to residents in the general public ('final demand') but to other business addresses (i.e. 'intermediate demand'). Producer services include professional firms supplying financial or legal services to other places of business, or more recent specialist activities such as computer services or market research provided for other employers. These purchasers may be in agriculture, mining, manufacturing, government or the rest of the service sector. Strictly speaking, a definition that took in all the suppliers of services to other businesses would include freight transport, wholesale distribution, security protection and canteens, all of which have also in fact been part of the trend towards manufacturing firms sub-contracting work to other firms (Marshall *et al.*,1988). In this chapter, however, I shall exclude those three and similar activities and concentrate on 'financial and business services', including insurance and banking as well as most professional and business services (Division 8 of the international NACE and UK industrial classifications).

The New Economic Geography really does need to focus on people like business lawyers or manpower consultants. Producer services are now part of the infrastructure needed by cities to compete with each other across North America or Europe in their own right and in attracting other kinds of investment. They have also been the biggest source of direct job growth, on the above definitions, and have therefore been seen as the engine of modern metropolitan change. In the USA, their employees increased in number from 5.5 million in 1971 to 13.2 in 1991, an addition of 140 per cent, with women gaining the majority of new jobs with an increase of 278 per cent (but not the highest pay). In the UK and France the number of jobs in this sector doubled over the same period. Figure 7.1 can cover only the shorter period from 1986 to 1994, in which 50 per cent increases in the UK and France were overtaken by higher job growth as the sector spread out to the Benelux countries and Southern Europe. By 1994, taking business and financial services together, the sector was responsible for more than a fifth of all jobs in the UK, The Netherlands, Spain and France, and an even

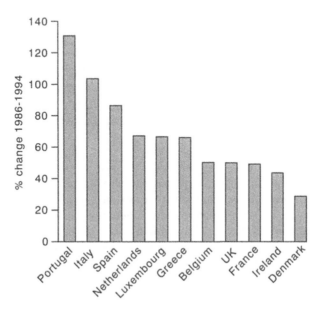

Figure 7.1 Percentage change in financial and business services employment, 1986–94
Source: CEC, Eurostat, 1993c, 1996c

133

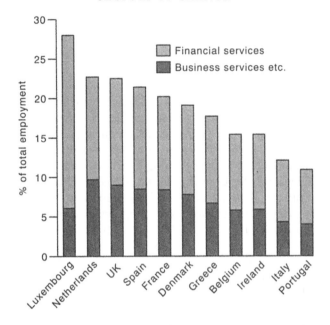

Figure 7.2 Dependence on employment in producer services, 1994
Source: CEC, Eurostat, 1996c

higher proportion in the internationally oriented small state of Luxembourg (Figure 7.2).

This widespread 1980s growth at a time of deindustrialisation in the wider workforce would appear fundamental to modern capitalism. A simple explanation for this change would be that manufacturing industry, in seeking flexibility, was sub-contracting work to producer services firms. On this basis, this rapid period of change might be a major one-off transfer effect of the 1980s. Geographically, it might also prompt suppliers to locate in offices providing close personal contact with new customers. Both of these points are valuable starting points for analysis, but they are only part of the pattern. Lentnek *et al.* (1994) argue that less than a third of recent producer services growth can be unequivocally attributed to this vertical disintegration. The principal point seems to be that the new realms of specialisation provided by new divisions of labour in finance and business services have added greatly to the productivity of the overall commercial system. It is also the case that people's unearned income from financial investments has increased relative to their income from employment, and that the demand for financial services appears to have risen disproportionately in high income countries. Van Ritbergen (1994) shows how insurance premiums in Western Europe increased as a proportion of GNP from 3.2 per cent in 1970 to 5.7 in 1991, with figures of 10.7 in the UK and 1.7 in Greece revealing a marked north–south contrast remaining at the end of the period. It was also notable that the European share of the world market had increased from 18.4 to 33.3 per cent.

THE GEOGRAPHICAL BASIS OF PRODUCER SERVICES' TRADE

Patterns of spatial concentration have traditionally attended the dynamic of producer services, whether between continents, nations, regions or metropolitan areas. International firms and their suppliers are most evident in national capitals and smaller firms lower down the hierarchy of centres, leaving specialised industrial towns without many producer services. What does this say about their supposed dependence on manufacturing industry's demand? Before exploring the location of producer services' jobs it is necessary to clarify just what their markets are like, through an approach developed over the last ten years (Harrington *et al.*, 1991).

The city of Edinburgh was ranked fourteenth among European cities for its business growth prospects by Cambridge Econometrics (1993). Its Planning staff asked me to investigate whether its producer services were

> an independent or a dependent source of growth. This should determine whether policy is directed towards encouraging such services or focused on other aspects of the economy on which they depend. This will have implications for property provision, enterprise support, training and international links.
>
> (Townsend and Macdonald, 1993: 444)

At the outset of the study it was clear that Edinburgh's large financial institutions, including General Accident and Scottish Widows, had UK and to some extent European and world markets for services. What was less clear was whether the rest of business services, all supplied by firms employing less than 400, had more than a local trade: there were reports of an Edinburgh estate agent expanding setting up in Prague and of a local firm moving into Silicon Valley in California (later confirmed), but were these typical?

A survey of thirty-nine business services firms in Edinburgh (83.0 per cent of possible cases in a stratified sample) showed their business to serve mainly the rest of the Edinburgh service sector. As in French studies, the role of *partners* in supplying a service was found to be important. There was a definite concentration of linkages within Edinburgh, but tailing out across Scotland and the UK, eventually to the international scene. The distribution of our firms' *sales* is best considered first by sector and then in turn by location. Were business services working for industry, say for the North Sea oil industry or manufacturing in central Scotland? Or were they, as Goe (1990) suggested for the USA, and Moulaert and Toedtling (1995) for Europe, working mainly for other parts of the service sector? A first clue to

135

this question came from the *sub-contracting of services to* our survey firms: although more than half the respondents had taken over work from their customers, little was from manufacturing, and most came from the privatisation of public sector work and outsourcing by local financial groups. In fact the sectoral distribution of sales extends across most of the Standard Industrial Classification, with 18.3 per cent going to the financial services sector and over 60 per cent to the service sector at large (Townsend and Macdonald, 1995). As seen here, manufacturing actually was exceeded in size as a market by the private household sector (which does not, of course, comprise 'producers' at all), while the general commercial property was prominent, for instance in work for leisure facilities, tourism and nursing homes.

The great majority of sales are thus to intermediate rather than manufacturing sectors, as affirmed in recent US work (O'hUallachain and Reid, 1991). This result may be affected by the high proportion of manufacturing branches in Scotland of foreign TNCs which may rely on the parent company for business services. (An American-owned subsidiary company would, however, be handled under Scottish law through Scottish lawyers.) It is nonetheless consistent with Scottish results of O'Farrell *et al.* (1993), showing that expansion of demand is the primary cause of increasing business service output, rather than industrial externalisation during restructuring.

More than 40 per cent of sales took place within the boundaries of the city. Remarkably, this is the mirror image of the pattern for Glasgow, which has 42.9 per cent of sales in the city itself, 18.8 across western Scotland but also sales across Scotland as a whole. This double test showed just under 20 per cent of sales proceeding beyond Scotland in both cases. Despite recent efforts to expand into England, and many continuing schemes of the EU to open up services trade between member states, the evidence of international trade to date is disappointing. Exports beyond the UK provide about the same proportion as in O'Farrell *et al.*'s (1992) sample of firms in Scotland, but well below the proportion of a fifth from their matching sample in South-East England. This was a difference that they attributed not merely to distance, but also to the restraints imposed by the less demanding and less complex nature of the existing market in Scotland.

International trade in producer services?

In a further study, O'Farrell *et al.* (1994) found that only low proportions of business service firms viewed their entry into international markets as a fundamental part of their corporate strategy. However, London firms were more deeply entrenched in foreign markets than other British ones because of the quality and specialisation of services they provided, and the strength of their London base among global TNCs.

If we compare our results with international data, we find the same contrast between primate and secondary cities. Table 7.1 shows how in a survey

Table 7.1 Distribution of sales from sampled firms in business services, Edinburgh (per cent)

	Parent city	*Region*	*National*	*International*
EU primate cities	32	19	38	11
EU secondary cities	29	27	39	5
35 cases in Edinburgh, 1993	42	20[1]	31[2]	7

Note: Data are weighted by employment.
1 East Central Scotland, excluding Edinburgh
2 UK excluding East Central Scotland
Source: Daniels *et al.*, 1991; Townsend and Macdonald, 1993

of 292 respondents, 11 per cent of sales from firms in capital cities were made in other countries, compared with 5 per cent in secondary cities. If Edinburgh were regarded as a secondary city, then our reports show it to be performing slightly better than expectations. Despite the increasingly global reach of London in the 1980s, and the EU efforts, the day of extensive international trade in services has not yet arrived. In Europe, 'the majority of the insurance companies are still locally run businesses, in spite of the fact that mergers, joint ventures and strategic alliances are daily news in this sector' (Van Ritbergen, 1994: 266). Moulaert and Toedtling (1995) argue that markets for producer services are mainly regional in most European countries, though wider in the UK, France and the Nordic countries.

For the future, Morgan (1992) considered whether personal financial services would be amenable to European trade, and suggested that the optimistic conclusions of the EU's Cecchini Report (1988) would turn out to be correct in the long run. This was because globalising trends in the provision of personal financial services would provide great increases in efficiency. Japanese banks are now entering the European city and North American producer services are expanding from their home regions into foreign markets, for example across the Pacific Ocean. If we look at sales beyond their home regions by business services (with inevitably varying definitions of region and the sector), we find that 24 per cent of producer services sales from Cleveland, Ohio passed beyond its region (Goe, 1990), 27 per cent from Seattle, Washington State (Beyers *et al.*, 1986) and 34 per cent from Vancouver, Canada (Davis and Hutton, 1991). In these two west coast areas, business has expanded well beyond the concerns of their traditional hinterland and involved items like software sales to the Far East.

However, we must note that the growth of services exports from metropolitan areas may only have served to offset the loss of manufactured exports lost during their previous deindustrialisation (Persky and Wiewel, 1994) and may not necessarily have increased overall exports from the individual city. Some firms interviewed in Edinburgh preferred to base their future trade on reliable local activities. As in Harrington *et al.*'s (1991) work, the activities with notable exports are the exceptions, i.e. particularly large, particularly specialised or parts of multi-regional enterprises. Nonetheless, a very significant part of the activity and jobs of cities is now heavily dependent on 'producer services' with an increasing spread of markets.

LOCATION THEORY FOR PRODUCER SERVICES?

The range of sectoral and geographical destinations for producer services sales makes their location part of a complex web of interaction between supply and demand elements, with the physical transport costs of industrial location theory replaced by a number of contingent human relationships in socio-economics. First, the concept of 'contact potential' reminds us of the critical importance of businessmen's personal travel time, and of nodes in the time-distance map, which make centres such as Stockholm, London or Dublin so important on the office location map of their respective countries. Second, it is the actual relationship between customer and supplier that must be considered, the 'relational necessity' of Monnoyer and Philippe (1991). Third, the arrangement of contracts must be seen as part of the network of contacts of the respective firms (Strambach, 1994), again having social as well as economic characteristics. Fourth, the opportunity for contracts must in turn be seen in terms of the theory of organisation of large corporations, with their need for a hierarchy of co-ordination of different functions, such as finance, investment and planning; purchasers must enjoy close interaction through face-to-face communication: 'this explains the tendency for Advanced Producer Services suppliers to locate in spaces where there is a relatively high concentration of corporate headquarters: metropolitan centres or large urban centres of national significance' (Daniels and Moulaert, 1991).

Gottmann (1961) emphasised how the office sector in the urbanised North-East USA (his 'Megalopolis') benefited from 'agglomeration economies' including the presence of universities, leading world libraries, and R&D facilities. These strengthen the tendency to concentrate skilled human capital, which Illeris (1991b) sees as the most important local condition for the strong growth of services. Cuadrado-Roura and Gomez (1992) argue from analyses of Spain, in comparison with other European countries, that supply factors confer high advantages on particular metropolitan cities. A large labour market of skilled people is a crucial advantage, which in the case of the City of London generates its own sense of culture and power as well as its own support staff (Allen and Pryke, 1994). Many people prefer to live in large cities, but clearly other areas attract people because of their social or physical environment. 'Old university towns are examples, and so are the "sunbelts" of southern France, Germany and the United States' (Illeris, 1991b: 105).

The play of structural factors is contingent on a wide range of human factors in the evolution of producer services' locations generally (Daniels and Moulaert, 1991). In practice, producer services have their own hierarchies, extending from an international merchant bank on the one hand to a local market-day only outlet in a Welsh hill town on the other. It has been found profitable to distinguish different kinds and levels of customer contact, because they impose different kinds of locational requirement. It was certainly possible in Edinburgh to find firms claiming they had deep roots in the 'New Town' because they worked together regularly in partnership with other firms there in the 'constructional professions', bringing together the different skills of architect, civil engineer, surveyor and estate agent on the same project; spatial proximity meant they could readily walk to joint project meetings. Tordoir (1994) argues that the simple idea of frequent personal interaction requiring spatial proximity can break down in different ways. A 'sparring' relationship of interpersonal and reciprocal interaction between customer and client certainly requires regular access to the client's premises at a high level. However, even tailor-made 'jobbing' work can be done entirely in the supplier's premises after initial meetings.

Beyond this, however, lies the widely recognised concept of the 'back office', in which a firm or government department may establish a large unit for more labour-intensive or routine work, such as clerical processing or computer operations. This choice is the occasion of recent debate about the dynamics of the office sector, for instance whether it has a part to play in the revival of rural areas.

LOCATION TENDENCIES AMONG PRODUCER SERVICES JOBS

In Europe there has undoubtedly been some geographical change attending the rapid growth of producer services. In reviewing 'business services in the core-area of the European Union', Van Dinteren and Meuwissen (1994) saw the main centres of jobs in 'the old economic core' of Frankfurt, Paris, Amsterdam, London and the Ruhr. Illeris (1991b) noted that the most rapidly growing services are overconcentrated in metropolitan areas, *but* that all kinds of services, including the generally slow growing, grew *less* there than elsewhere, producing a marked shift in the location of jobs in total. In fact, Paris has been showing a net loss of jobs in this sector to the surrounding Ile de France since 1968. So too has there been more regional dispersal in Germany and Belgium, for instance to Cologne or Antwerp. Van Dinteren and Meuwissen (1994: 366) conclude that 'employment growth in business services has been relatively higher in regions adjacent to metropolitan regions than in the metropolitan areas themselves'.

In understanding the process of locational dispersal, in connection with contacts with customers, networks of partners and the corporate hierarchy, it is necessary to recognise both the speed of employment growth and the frequency of relocation by business service firms. Thus, in his survey of the Randstad, Holland area, Hessels (1994) noted that 50 per cent of firms interviewed had relocated at some time in the past. Push factors related to accommodation lay at the root of moves of decentralisation, leading, as in North American suburbs, to the growth of subsidiary centres for business services, although new firms were being founded too in these areas. The point of most important theoretical interest was that linkages were not necessarily local but that suburban firms were more oriented than the average to extensive national markets, reached by motor vehicle. The policy implication of this is that, if, as they do, producer services offices act like 'export factories' in the local economy, then supportive town planning policies will not necessarily want to concentrate entirely on city centre districts.

Moulaert and Toedtling (1995) concluded a thorough review with the concept that the European capital cities were certainly losing out relatively to the remainder of their countries, but that there was no clear pattern of which type of place were the winners from the process of deconcentration of business services. This contrasts with the continuing heavy concentration in metropolitan areas in the north-eastern seaboard of the USA. Despite the regional shifts in the US economy, these services and the corporate headquarters that they serve are still heavily concentrated in the seventeen cities of more than 2 million people. Beyers (1991: 170–1) found that 'producer services have exhibited negligible deconcentration from metropolitan areas over the past decade, but there has been a modest redistribution of work between areas'.

PROCESSES OF DECENTRALISATION AS EVIDENCED BY SOUTH-EAST ENGLAND

A metropolitan tendency is well-evidenced by Europe's biggest office centre. Virtually half of the UK's production (by value) of financial and business services is undertaken in one region, the South East; in terms of UK employment at end-1995, 44.8 per cent of their employees worked in that region – 25.0 per cent in Greater London (GL) alone – compared with 32.5 per cent of employees in all sectors (*Labour Market Trends*). In turn, producer services accounted for virtually a third of total output in the South East. In business services an even greater concentration, biased toward the west and north of the region, speaks of the importance of 'centrality within southern Britain space-economy for reasons of access not only to London but also to clients nationwide' (Keeble *et al.*, 1991: 450).

The metropolitan region of South-East England has repeatedly evidenced the directions of change that we have described. As Illeris (1994) indicated, the most important, dynamic and well-paid activities remain in or near the city of London; yet the South East beyond the boundaries of Greater

London (GL) now has almost as many financial and business services jobs as GL (19.7 per cent of the national total). Further office relocation from London has overspilled its rather arbitrary boundaries, easily reaching Bristol, Northampton, Peterborough and Norwich. Our central questions in this section will be to ask what is the balance of relocation (from London) and of indigenous expansion in this growth, and to what extent do these dispersed activities still rely on contact with London, perhaps as 'back offices' located in virgin areas to secure cheap secretarial labour (Van Dinteren and Meuwissen, 1994; McDowell and Court, 1994b)? The issue can be compared to the relocation of manufacturing industry to areas of more plentiful women's labour at many dates before and after the Second World War.

In practice, office dispersal has never been as easy to achieve as the long-distance movement of manufacturing, and government intervention is needed if market trends are to be offset (Martinelli, 1991). Although manufacturing employment in Inner London began to decline as early as the 1950s, office employment grew, and burgeoned in the early 1960s. Thus the fitting of new offices into the London map and skyline became a big problem, as did the increased numbers of rail commuters on the suburban and wider network. The government then not only established a programme for the dispersal of its own expanding work, which was successful in terms of numbers of back office jobs relocated, but also set up the Location of Offices Bureau (1963–79) to encourage firms to leave London. It was hoped that, like factories under regional policy, offices could be induced to disperse well beyond the boundaries of the South East to most regions of the UK.

After this late start, there were two main periods of dispersal from central London, one from 1965 to 1976 and another starting fairly late in the 'Thatcher boom' from 1986 to 1990. In the first period, firms' many attachments to the metropolitan region constrained most moves which were to sub-centres of GL, or to towns such as Watford or Brighton in the Rest of the South East ('ROSE'). While factories needed only a few key managers and workers to transfer to a new branch, it was found that 25 to 40 per cent of employees were needed to transfer with an office to secure its continued viability, involving many issues for the resettlement of spouses and families of transferred staff (Bateman and Burtenshaw, 1979; Sidwell, 1979). In the conditions of those days, office messengers frequently carried papers between London and the new office, generally restricting relocation distances for the new branches to two hours' travelling time.

Improved communications in the 1980s clearly did not obviate the need for face-to-face meetings in London. An office development boom in the City and the Canary Wharf area of the London Docklands was 'almost wholly unexpected' (Diamond, 1991) but attributable to the inter-related IT revolution and the international expansion of producer services. These were aided by the use of very large sites in the City, the efforts of the Docklands Development Corporation and the general climate of investment. Taking

Table 7.2 Decentralisation of financial and business services
in South-East England (employees, thousands)

	Greater London	ROSE	Total
1981	568	276 (32.7%)	844
1986	694	391 (36.0%)	1,085
1990	808	540 (40.1%)	1,348
1992	699	521 (42.7%)	1,220
1995	739	540 (42.2%)	1,279
Change, 1981–95	+30.1%	+95.6%	+51.5%

Source: Department of Employment, Quarterly Estimates, via
National Online Manpower Information System

GL as a whole, employment in financial and business services expanded by
nearly a quarter of a million between 1981 and 1990 (Table 7.2), particularly
in the newest and most miscellaneous business services, followed by banking
and computer services. Over 100,000 jobs were then lost by June 1992 in the
heavy recession that followed, although it is important to note that job levels
then slowly recovered – and with them the numbers of commuters into
central London.

These data are, however, somewhat deceptive. Inner London was losing
jobs relatively to the rest of London (growing over twice as fast) and to
ROSE. The city's importance was taking on a different character. On British
evidence, Marshall (1994) argues that corporate hierarchies are undergoing
substantial reorganisation. Employment in head offices is declining, being
'hollowed out' as growing demand for business services is being supplied
by other firms. 'As they grow, large metropolitan areas are increasingly
becoming the base for higher order managerial, business service and financial
services, with routine administration being relocated to their hinterland'
(Marshall, 1994: 41). This decentralisation means that the hinterland of the
capital is emerging as the administrative support region for the large firm.

Thus, to generalise Table 7.2 slightly, financial and business services'
employees from 1981 to 1995 expanded by about a quarter in GL and a half
in the South-East as a whole, due to *doubling* in numbers in ROSE. Figure 7.3
covers the longer period of 1971 to 1994.

Expansion in the Home Counties – the case of East Surrey

What processes lay behind the doubling of producer services jobs in the
Home Counties? This is important because it has led to traffic jams in
previously quiet town centres and leafy lanes, and socially reconstructed
space in what had sometimes been exclusively commuting areas. To take just
one example from the North Downs, the main IT Centre of the insurance
firm Legal and General, established near the branch station of Kingswood,

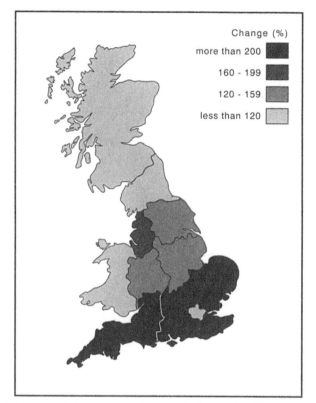

Figure 7.3 Regional change in employment in financial and business services,
GB, 1971–94
Source: *Employment Gazette* (various dates)

Surrey, and, looking like a very large crematorium, employs 2,200 people at an average salary of £15,000 per annum, thus disbursing £33 million (before tax) across its travel-to-work area.

What are the new processes of change, compared with the earlier period of movement promoted by the Location of Offices Bureau? A survey was undertaken by the author in Surrey in September 1992. With a mere 85.5 per cent increase in financial and business services from 1981 to 1991, this area barely exceeded the ROSE average of 80.9, far below counties in the 'Western Arc' around London. The survey was conducted in the three easternmost districts of Surrey, simply among the eighteen establishments from this sector employing more than 100; of them fourteen replied, covering 7,400 jobs or nearly a third of the sector's total employment in the area. This meant ignoring the very considerable growth of small business in the area, one of 78 per cent over ten years.

Are these simple appendages of London headquarters following the

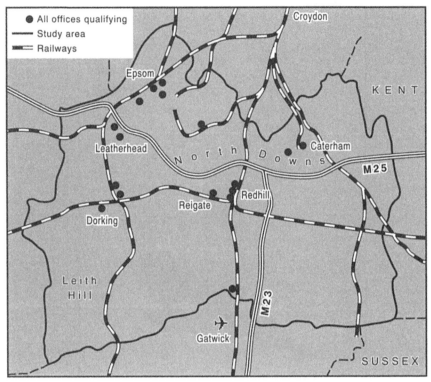

Figure 7.4 An example of 'Home Counties' growth in GB; the East Surrey
Study Area

branch plant model? The map in Figure 7.4 provides some clue, in that many
of the qualifying offices are close to suburban stations on routes to London's
Victoria or London Bridge termini; for example, a number of civil engineering firms in premises near Victoria lost their premises together in
redevelopment and chose full relocations on these routes in order to retain
staff. But there was no single model of evolution. Surveyed firms had been
moving out in this direction since 1939, some of them in 'stepwise radial
movement' in two or more stages, involving prior intermediate locations in
Kingston, Wimbledon or Croydon. Thus our population of large buildings
of 1992 was established gradually, though accelerating in floor area from
1986 to 1989 with building replacements and extensions rather than fresh
relocation from London.

Although the insurance firms retained head offices in London, more than
half of the premises surveyed had become the head offices. All had national
functions, supported by subordinate regional offices. Some serviced national
manufacturing firms, notably in providing car insurance, hire purchase and
market research, but most work was for other services, and exports were

again low, at about 5 per cent of total activity. In the majority of cases, large domestic customers, agents and regional offices could log on to the computer system maintained in East Surrey. Some of these offices had been among the pioneers of computing in the 1960s and now regarded data bases as the life blood of calculating transactions and of storing millions of policies (though this did not necessarily provide in all cases for larger or more complex tasks to be undertaken).

The 1980s expansion in East Surrey was thus based on large-scale computing; although originating as back offices in the 1960s, several had become freestanding national entities and had expanded on adjacent sites – or in other cases, under a change of national Town Planning requirements, around Redhill station. These results both compare and contrast with a previous colonisation of Surrey, by research laboratories. These resulted principally from a search for premises in reach of London for national levels of work. Although the environmental advantages of Surrey were appreciable, the area was of little account for particular linkages with other establishments, which again developed only incidentally.

A more radical view of Home Counties re-location?

Thrift *et al.*'s (1987) 'Sexy greedy, the geography of the new international financial system' epitomised the high salaries, excess consumption and borrowing of the South-East Service Class in the 1980s. In this context, Surrey shared with GL the highest household disposable income per head, 20 per cent above the national average in 1986, and established a higher income by 1991. It is, however, low pay and exploitation, especially of women, that excites critical attention among geographers. McDowell and Court (1994a) argue that back-office decentralisation appears to be a gendered process of cost reduction, with firms pushed out from central London by the high cost of both staff and premises in search of women's cheaper clerical labour and part-time work.

Sassen (1991) argued that the *Global City*, London being one, was developing a polarised income structure because producer services and related sectors were generating both some very high incomes and a prominent sector of part-time and casualised jobs at the bottom end of the income distribution. Could it possibly be that Surrey too was being polarised, that its high men's commuter incomes were parallelled by exploitative local jobs for women? Have financial and business services used computers to create a de-skilled and dual society? Are back offices exploiting the lower skilled in the spatial division of labour, avoiding training and a trades union presence? Are they too heavily dependent on a close London link, and disproportionately vulnerable to recession? Do they represent examples of sub-contracting and the use of a 'core and periphery workforce'?

The fourteen establishments reported that they would be much more labour-intensive in their operations if it were not for computerisation. Their clerical component would otherwise be twice as great, and the proportions of technical and professional staff much lower. The ROSE area shows average earnings levels in this activity a quarter lower than Greater London for men and a fifth lower for women (*New Earnings Survey*), whereas surveyed firms claimed to be paying only 5 to 10 per cent less than central London. Among the fourteen establishments, the median managerial and professional workers were paid an average of £29,667 p.a. (slightly above the GL average for white collar men in the sector), while the clerical and secretarial average of £12,375 p.a. was slightly below the GL average for women white collar workers.

Surrey establishments reported that availability of labour had been very sensitive to salary adjustments, the power of the central London labour market and the risk of losing people to the intermediate centre of Croydon. Replying in 1992 when the 'Thatcher boom' up to mid-1990 was still a recent memory, the largest firm had studied the Census, scoured the county schools for leavers and devised schemes for women returners and twilight evening shifts. Several firms reported a tough recruitment problem before the last recession; they could not retain people, who were leaving for self-employment. The idea of abstracting London commuters was not often raised as a serious possibility, although it had worked at the beginning of the recession.

Due to the proportion of engineering and computing firms in East Surrey, they have a lower proportion of women workers (37.4 per cent) than the producer services for ROSE (53.1 per cent). Part-time proportions stood well below the regional average of 15.7 per cent in 1991. From 1981 to 1991 the numbers of part-time and women staff had increased somewhat faster than the average for the region.

There is then a narrow majority of women in the ROSE's financial and business services and the workforce is changing in the direction hypothesised for women's and part-time work. But, as in central London (McDowell and Court, 1994b), this process is very slow and there are few women managers save in personnel. The caricature of polarisation is, however, represented only by the extreme of our fourteen firms. The highest paying firm, an elite group of actuaries, paid an average of £40,000 p.a. (plus bonuses), while the two lowest paying establishments were dominated by clerical and secretarial workers paid in one case £6–8,000 p.a., and in the other £6,500–14,000 (averaging £9,000). Interestingly, these two were the most dependent on women staff, to the levels of 60 and 65 per cent, the most locally recruited of the survey. One, a market research firm, saw data input work as a low paid, non-career women's activity, while the other provided routinised insurance work for old people and was described as a 'sausage machine factory' by the personnel manager herself.

These two, but only these two, do support the hypothesis of the low paid back office. Beyond that, all but the two largest firms reject having unions, or even a staff association. The firms are benefiting from 'flexible special-isation' insofar that many had taken in work from national customers, previously undertaken by them, partly as a result of the government's deregulation of financial services, two having taken over car insurance or hire purchase from car distribution companies, three having provided more planning, advice and administration for clients' pension schemes. There was little sub-contracting from our establishments, except for cleaning, catering and security, although in-house programmers were commonly on self-employed contracts. There was little sense of community surrounding their Surrey premises, even when leisure and sports facilities were attached, and the conclusion was that the firms lacked strong attachments to the county.

'Could the work have been located in northern England?' This question prompted firms to repeat familiar reasons, that they were dependent on get-ting business from London, and that such a move would have lost them too many staff. Two insurance companies were well aware that present day com-puter communications would allow moves that were impossible when they chose a Surrey location in the 1960s. The great majority realised the potential locational mobility provided by modern telecommunications and had been spurred to look at the question by the property costs and labour shortages building up before the recession. Two had talked of 'moving north', to Oxford and Peterborough (sic), but one had undertaken a feasibility study in North-East England, and three had moved work, or in one case considered moving it, to associate offices in the North. Politically, then, a different gov-ernment regime could have had controls which might have shifted some of these expansions beyond the South East, during or preferably before the 'Thatcher boom', as did tend to occur spontaneously in manufacturing in the 1981–9 period (Townsend, 1994b). In the next section, we consider what growth occurred in any case outside the South East. Will growth continue to concentrate in metropolitan regions, now broadly defined?

GROWTH IN NON-METROPOLITAN AREAS? INTERNATIONAL COMPARISONS

The fear from the first office location studies in the late 1960s was that provincial centres would persistently lose out to the centralising tendencies of corporate mergers and new specialist functions. Recession conditions of the early 1980s, together with the subsequent recovery, appeared to under-write the relative importance of the South. The argument of Leyshon and Thrift (1989) that 'South goes north', through the revival of the British provincial financial centres, was certainly refreshing. Centres such as Leeds or Manchester were growing, through the use of online facilities to the London Stock Exchange, the development of fuller regional networks of offices by national firms and the externalisation of tasks by regional industry, not least in handling the accountancy of deindustrialisation.

Table 7.3 The North–South divide in financial and business
services, GB (employees, thousands)

	North	South	Great Britain
1971	487	831 (65.3%)	1,318
1981	640	1,091 (63.0%)	1,731
1990	944	1,757 (65.0%)	2,701
1992	981	1,611 (65.0%)	2,592
1994	1,011	1,715 (62.9%)	2,726
1995	1,028	1,719 (62.6%)	2,747
Change, 1971–95	+111.1%	+106.9%	+108.4%

Note: South comprises the South East, South West, East Anglia and the
East Midlands; the North comprises the rest of Great Britain.
Source: Department of Employment, Quarterly Estimates, via National
Online Manpower Information System

Taking, however, one set of statistics (Table 7.3) and including the
South West (with decentralisation to Bristol and Bournemouth), the East
Midlands (including Northampton) and East Anglia (with Peterborough
and Norwich) in the 'South', it is remarkable that the share of the South
was merely oscillating cyclically within a narrow range over twenty-one
recorded years, happening to show a slight net decline over the years 1971–
95. The North certainly more than doubled its total of jobs, but so too did
the South. When we look at individual metropolitan areas for the period
1981 to 1991, we find that Edinburgh contrasts with the larger provincial
cities, which have an average growth rate below that of GB at 33.5 per cent
and are all in the 'North'. It is true that producer services shunned many of
the areas of more purely industrial background, although the growth of
long-distance telephone sales was generating clerical and similar posts in
large offices in, for instance, Leeds and Newcastle-upon-Tyne, in new
banking arrangements, airline ticketing and other activities (Marshall and
Richardson, 1996).

I intend this extensive focus on research results from the UK as a basic
insight into processes that may be international. To extend the review
abroad, we find that non-metropolitan but prosperous parts of Europe such
as southern Bavaria, French-speaking Switzerland and south-eastern France
bear witness to producer services growth, whereas areas that have depended
on branch factories for manufacturing work have, as in Britain, failed to
generate much significant activity in producer services. Martinelli (1991),
working on Southern Italy, reaches many of the same conclusions as Marshall (1985) over the relationship of business services and regional policy.
The predominance of branch plants among the modern industrial developments of Southern Italy tends to mean that purchases are organised from the
firms' headquarters in other regions; demand is poorer and more variable

from indigenous firms; the overall result is that service employment totals are more dominated by public service work than in other areas. European authors studying uneven development, such as Marshall and Jaeger (1990), have also noticed how producer services tend to shun towns with a purely industrial history. Thus Van Dinteren and Meuwissen (1994) note that the worst prospects for producer services in the heart of Europe lie in Wallonia (Liege and Namur), Metz, Nancy, Saarbrucken and the Ruhr. On a wider scale

> London, New York, Tokyo, Paris, and Sydney are attracting new pro-
> ducer services or experiencing the expansion of existing services at a
> much higher rate than Liverpool, Detroit, Osaka, Lyons or Newcastle
> (New South Wales).
>
> (Daniels, 1991: 136)

The movement of these services to outer rural areas is proceeding but is not outstanding in terms of volume of jobs or rates of growth. In North America, Kassab and Porterfield (1991) note that recent research downplays the role of producer services in rural development. Holding constant the type of service activity, they report a tendency for larger proportions of non-metropolitan workers to be employed in lower level positions. However, they found that the greatest disparities between metropolitan and non-metropolitan employees lay in the greater prevalence of part-time work and the lack of company health insurance. Thus we may conclude that there are qualitative as well as quantitative differences between metropolitan and other areas in producer services; decentralisation has proceeded less than in other sectors, and the highest paid men's jobs have not moved far at all.

COMPUTER SERVICES – ANY DIFFERENT?

Do the expanding parts of the sector indicate any shift in directions of change? Does this leading growth sector behave differently from others, in relation to metropolitan and other locations. Gentle and Howells (1994) estimated that European computer services had attained a workforce of 350,000 employees by 1989, but that 90 per cent of its markets were still domestic to individual countries.

In the UK, the largest supplier in Europe, the Computer Services Association estimates that foreign trade was somewhat higher, at 15 per cent, but that figure may be weighted toward their membership among larger firms. In the UK, employment increased from 54,800 in 1981 to 137,800 in 1989 and 170,900 in 1994 (*Employment Gazette*), increasing at nearly five times the speed of financial and business services in general. This reflects the widespread growth in importance of IT, the swing toward the market in software rather than hardware, and the salience of independent computer services firms and establishments. A major feature has been the growth in the provision to other companies of 'facilities management' and of 'complete business solutions' incorporating both hardware and software together.

Coe (1996) reports that, as shown by the distribution of employment, the benefits of growth in computer services have not been evenly spread. The regional picture in 1991 was dominated by the South East, with 56 per cent of GB jobs, a proportion which, as in the wider sector described above, was stable over the period 1981 to 1991. Compared with producer services as a whole, what was different was that the respective shares of GL and ROSE were completely reversed, with the latter having a full third of the national total by 1991, and GL just 23.2 per cent. Further data confirm the strong position of the 'greater South-East' counties, particularly in a 'western arc' (Hall *et al.,* 1989) around London from Hampshire to Cambridgeshire. At the other extreme, 83 out of the 322 statistical units of GB had no representation in this sector at all. With the exception of three North-Western units, Warrington, a New Town, and Northwich and Macclesfield, which provide a social environment not unlike the South East within easy reach of Manchester Airport, almost all the units outside the 'South' of England contain a large population centre. Coe (1996) agrees with both Kelly (1987) and Cooke *et al.*(1992) that computer services tend more to urban locations than the rest of producer services, due to their orientation to markets through personal contacts, although some sub-sectors such as consultancy and software design may cluster in residentially attractive suburban locations. Overall, this data would seem to suggest the kind of regionalised mode of service growth across the South which is then somewhat 'dislocated' from the rest of the UK, as suggested by Allen (1992).

Coe went on to assess some of the processes involved through comparative interviews undertaken in Hertfordshire and Tyne and Wear. The results highlight the differential processes at work in creating the divergence between a metropolitan and an old industrial region. The activity in Tyne and Wear is restrained largely by structural conditions, i.e. by dependence on a somewhat retarded branch form of manufacturing, by the computer industry's historical development in the South and the overall concentration of corporate headquarters and the service economy in that area. Unfortunately for the North East, the improving telecommunications and network services available do not appear to be reducing the need to be geographically close to clients in supplying many services; around 50 per cent of both samples had remote links into their clients' computer systems, but in only a few companies was this seen as a major part of system support, often due to fears of losing confidential information. 'The overriding importance of face to face contact (in most subsectors) came through repeatedly in interviews, both in terms of customers' requirements, and in terms of a firm's perception of its quality of service' (Coe, 1996: 75).

WARNINGS FROM RECESSION

'De-tertiarisation' began to appear as a term to reflect redundancy and job losses in the service sector during international recession conditions of 1990–2. According to Sjoholt (1994), job losses began to affect the Scandinavian service sector, especially in metropolitan areas, after several decades of continuous growth, while reports of computer services' decentralisation to Ireland, India and Singapore became more persistent. Three of the Surrey firms launched into tirades about having been restructured after foreign takeovers. 'The 1980s culture when all we had to do was write the business, and the profits would look after themselves, is gone forever.' Six of the fourteen establishments reported formal redundancies in the recession, and all but two had reduced employment since 1989; those that were still in profit were rationalising their operations or restricting recruitment. The average reduction there was about 10 per cent, and that stands as a fair description of the outcome in producer services in the South East generally.

Although better placed sectors and geographical areas for growth had offset these losses fairly quickly, it is argued that the recession merely exposed longer-term weaknesses (Sinden, 1995). Certainly the mix of growth within financial and business services in Britain appeared to change. For instance, the mantle for leading growth was taken from computer services by new and specialist services. Undoubtedly, 'banking and finance' itself fell into unprecedented decline as regards employment, falling from 625,500 in 1989 to 549,000 in 1995. Massive job losses reported regularly from all the main 'retail' banks are argued by Sinden (1995) to have resulted from the interaction of deregulation, technological advancement and increased competition, together with rising corporate pressure to maximise profits. Work organisation in these banks is becoming more flexible, but individual jobs are becoming more rigidly defined and constrained than previously. Job losses were heavily concentrated in South-East England, resulting from changing work practices and the decentralisation of computer centres. However, data for the period 1991 to 1994 suggest that the percentage reduction was approximately the same in the South East and Britain as a whole.

The financial world has discounted these changes as temporary effects of recession. They see the further growth of financial services restricted by job-shedding tendencies, notably in ordinary 'retail' banks, producing a polarisation of careers in the 'flexibilisation' of working practices, the further 'feminisation' of financial services employment, a reduced rate of decentralisation from London, and sharp movements of jobs between travel-to-work areas as firms rationalise their activities and seek buildings suitable for the installation of IT.

The net effect of these countervailing changes has been for producer services to lose the lead in forecasts of employment growth. Thus, Lindley and Wilson (1993) foresee structural change being determined mainly by shifts within the service sector, with many activities related to leisure, tourism, health and education showing a greater volume of employment growth than in financial and business services. They also see the rationalisation of banking and insurance jobs as further restricting the prospects of women in the sector, after their jobs had in fact grown very little from 1991 to 1994. These forecasts have to be seen in the context of international developments.

There is the argument that Britain must certainly join the European Currency Union if London is to retain its profitable supremacy in European financial affairs, but that in any case 'everybody in Paris and Frankfurt is concentrating on breaking London's domination of the equities market' (Whitebloom and Springett, 1994: 38). Beyond the European scene, the growth of certain Third World cities to 'World City' status is a possibility. One competitive threat to rich countries' cities is from newly-industrialising countries which are increasingly providing quality software services at extremely low cost, often supported by government policy through import restrictions and incentives. Thus, Singapore has encouraged the establishment of software centres by Unisys and Hewlett-Packard. Similarly in India, a state-owned company has won contracts for providing low cost software for London Underground.

On the other hand, however, the harmonisation of service trade through WTO and EU agreements has meant that it is difficult for governments to favour their own national firms. Future change might only help First World countries export to Third, thus further marginalising the poorer countries. Although some producer services TNCs will always follow their fellow manufacturing firms, the number of examples like the growth of Indian software exports 'are both sectorally and geographically limited' (Gentle and Howells, 1994: 320). It may be in any case that employment growth of producer services is nearing its peak in the countries that first adopted them. The advanced countries might be reaching saturation level (Moulaert and Toedtling, 1995).

SUMMARY

- The growth of activity and employment in producer services is based on supplying the service sector at large, although the sub-contracting of work from manufacturing firms does contribute.
- Despite attempts of the EU to facilitate international trade in services, export trade is generally limited and regional trade quite considerable.
- There is a clear locational tendency for producer services to be strongly based on metropolitan centres, but for all but the best paid men's jobs now to be decentralising.
- The example shows that the scale of jobs in Surrey has been achieved over

a long period. There is an idea that the jobs are low paid, part time, women's jobs in back offices, but it was found to be wide of the mark.

- Computer services still depend heavily on face-to-face contact, and are heavily located in the 'western arc' around London.
- Job levels in some of these services, especially banking, may never recover from the recession of the early 1990s.

8

NEW FLEXIBILITY IN RETAILING

A wave of realisation has just swept Geography, that change in retailing is highly significant to culture, landscape, economy and society – not least in selling to women and employing women (Lowe and Wrigley, 1996). Retailing changes the way in which people make their livings even more than it creates new jobs. I begin with the changing nature of consumption, as shopping and leisure facilities become further merged. Across the EU, there is cultural variety in the practice and regulation of shops, but all tend to converge on the French or British style of hypermarket, distinctive in its use of the 'numerical flexibility' of part-time staff.

Retail corporations show a new concentration of power:

> Top retail firms are now run by polished professionals. They exert enormous power and influence over manufacturers and consumers – and over urban, suburban and rural environments the world over. Retailers have grown, first at home and then increasingly abroad, into some of the world's largest companies, rivalling or exceeding manu-facturers in terms of global stretch.
>
> (*The Economist*, 4 March 1995: 3)

In a period when producer services have gained acceptance as vital parts of the production economy, there were relatively few innovations in the inter-pretation of consumer services, despite the arrival of TNCs in retailing and tourism. I dismissed retailing, for instance, as static in employment terms (Champion and Townsend, 1990). Likewise

> While the culture of consumption has received increasing attention in the 1980s, consumption workplaces and production methods have been relatively neglected. One reason for this neglect is that the retail sector was considered peripheral to processes of industrial restructur-ing. It was one of the places where displaced manufacturing workers (or more likely their wives and children) were likely to land but, as a static sector, had little intrinsic interest.
>
> (Christopherson, 1996: 159)

THE CHANGING NATURE OF CONSUMPTION

Consumption of all kinds has increased so heavily in the industrialised world as to attract international capital to re-devise the nature, production and employment of services. Dawson and Burt (1989: 5) have described the 'steady rise in European consumer power' leading to large retailers, national and TNC, taking control over previous links through dominating their manufacturing suppliers, accentuating the need for flexibility in the whole supply chain (Foord *et al.*, 1992). The growth of sales has propelled employment upwards in most countries, but more significant has been the pace of competition which has transformed the work and hours of shopworkers through the process of flexibilisation.

> The 1980s were a particularly favourable period for the expansion of consumer services, fuelled by the growth of consumer expenditure as a proportion of national income, and a widespread increase in personal indebtedness. The late 1990s are unlikely to witness such a consumer boom, as the problems created by indebtedness in the 1980s (e.g. in property) take time to unravel. In such a climate the limits to the capacity of even the very largest retail scheme for employment will become more apparent, especially with employment losses elsewhere in local economies.
>
> (Marshall and Wood, 1995: 169)

The broad changes that have driven consumer activity in European society may be categorised as demographic, socio-economic and attitudinal. Alongside poverty has come a general growth in incomes for those in employment, especially in the growing number of dual-income households. New lifestyles of 'dinkies' and 'yuppies' have contributed to the development of specialist services. For others the growth of car ownership and foreign holidays adds to a feeling of mobility. In general the structure of personal expenditure has shifted from physical products (food, household equipment, etc.) to service-based expenditure on leisure, education and health-related items, notably in recovery from recession in 1995.

Retailing's links to leisure in general

Although tourism has sufficiently strong growth of its own to merit a separate chapter (which follows) it tends to enjoy strong links with shopping. In international resorts, shopping for luxuries and souvenirs is a major way of capturing visitor spending. On the other hand, shopping can be seen as a major motivation for day tourism. Thus, a full-scale British survey (Baty and Richards, 1991), covering 630 million leisure outings of more than three hours duration in 1988/9, found that non-routine shopping represented the purpose of 10 per cent of the trips, but accounted for no less than 28 per cent of expenditure, averaging £22.80 per person per trip.

Leisure shopping was thus contributing sizeable sums to the economies of cities (for example Birmingham, £140 million p.a.) and resorts (Blackpool, £49 million p.a.), with resulting employment gains. Equally, leisure shopping for luxuries was adding new types of outlets to shopping centres, which from the 1980s adopted earlier American ideas of integrating leisure facilities with shopping centres, trying to capture more of the population's greater discretionary income and time (Howard, 1990). 'The shopping centre has become a place in which retailers sell and customers shop, but it is more than that – it also provides for entertainment, edification, education and sustenance' (Goss, 1992: 159). Some of the addition of small leisure items to shopping centres does not really constitute commercial leisure at all, simply the creation of a pleasant environment for shopping. But the combination of retailing and leisure gains synergy from longer and more free-spending visits from a wider area, better marketing and economical sharing of a common infrastructure, as some shopping and retailing activities occur at complementary times.

Taken together, jobs in the distributive trades, hotels and similar activities in E10 expanded from 19.6 to 22.7 million between 1983 and 1992 (Table 8.1), an average rate of 15.6 per cent, exceeded most strongly in The Netherlands and Greece, but followed by net job losses in France, Italy and the UK up to 1994. This represents only about a third of the rate of expansion of producer services, and is only a crude figure which takes no account of differing hours of work. Nonetheless it provides a positive contribution to jobs: and retailing is important to the labour force not only for the number of jobs but for change in the nature of work.

Table 8.1 Employment trends in distributive trades and hotels, 1983–92, E10 (thousands)

	1983	1992	Change (nos.)	Change (%)
Belgium	632	671	+39	+6.2
Denmark	373	403	+30	+8.0
France	3,655	3,783	+128	+3.5
W. Germany	4,430	4,862	+432	+9.8
Greece	589	752	+163	+27.7
Ireland	197	221	+24	+12.2
Italy	4,141	4,430	+289	+7.0
Luxembourg	30	33	+3	+10.0
Netherlands	881	1,227	+346	+39.3
United Kingdom	4,678	5,193	+515	+11.0
Total, E10	19,606	21,575	+1,969	+10.0

Source: CEC, Eurostat, 1993a, 1994

INTERNATIONAL CHANGE IN RETAILING

Retailing is the exceptional sector, for the role of its small firms is every-where declining and the investment of capital more dominant (Ducatel and Blomley, 1990). Since the emergence of the department store and mass merchandising at the turn of the century, an increasing proportion of sales and employees in the US retail sector has been controlled by large corporations (Christopherson, 1996). Although smaller corporations and independent retailers still make up the majority of firms and establishments, the largest firms now have more than half the jobs.

Change in the geography of retailing may be understood only through the institutions involved (Burt and Dawson, 1990). Equally, they have had to show sensitive response to economic, social, technological and cultural developments in society, including the impact of recessions on the need for flexibility and adaptability. It is sometimes the saturation of markets that has led retailing firms to expand on a truly global basis (Laulajainen *et al.*, 1993), while UK retailers are increasingly recognising scope to move into Europe that they had hoped to find in North America (Alexander, 1995).

Chapter 5 asked whether Europe might be following the USA in its employment structure. The US record in retailing shows not only a decline in the importance of small independent businesses but also, up to 1982, a marked rise in flexibilisation; the proportion of part-time workers, mainly women, expanded from 32.5 per cent in 1972 to a peak of 38.6 per cent in 1983 (Christopherson, 1996, Table 8.5). However, in the late 1980s and into the 1990s the shift was away from part-time work. Christopherson attributed this shift partly to the removal of retail centres to out-of-town locations and partly to demographic constraints on labour supply. She judged that employers were therefore switching from a strategy of numerical to one of functional flexibility. In the early 1990s the retail sector appeared to be shedding jobs as part of a cost-cutting strategy.

In E10, similarly, employment expanded from 1985 to 1990, before then shrinking as recession set in from 1990 to 1994. Compared with many other aspects of retailing, its employees are somewhat taken for granted, and under-researched. There is comparatively little trade-motivated interest in public documentation of employment issues because of the general lack of labour supply problems in a regime of high unemployment. Outside retailing

> commentators on the inner city have generally ignored and largely overlooked the considerable importance of the CBD to the inner city; to its functions, especially retailing and commercial-service activities; and to its related employment and space requirements.
>
> (Rowley, 1993: 111)

Likewise, while everybody knows that the major growth of out-of-town food superstores disadvantages poorer social groups, many forget that problems of access cut off not only potential customers from shopping opportunities but also possible employees from jobs.

THE EUROPEAN EXPERIENCE

The 12 EC countries are currently in quite different situations with respect to this typical pattern. An interesting and important point for discussion is whether these differences are simply the result of different stages of "progress" (and therefore transitory) or whether they reflect specific national and cultural characteristics.

(Kruse *et al.*,1993: 19)

The structure of retailing reflects the cultural characteristics of the society it serves. All sociological, economic and technological changes have their impact on retail trade and organisation. One such factor is geography, evidenced by the role of difficult communication patterns in maintaining the traditional retail structure of Southern Europe, based on small, independent shops. The national regulatory regime plays a strong role, incorporating cultural and institutional factors, such as laws on hours of opening or job contracts.

Any overall scheme of European retail evolution, with all its implications for consumers and workers of the sector, must be an over-simplification. The greater demand for a variety of goods and services is influenced by the greater employment of women in other jobs and by the ageing of population, but above all by the strong growth of private household consumption as a whole, which increased by 27.1 per cent in E12 from 1980 to 1990 (with no country increasing by less than 14 per cent; CEC, Eurostat, 1993b). Employment in European retailing has grown more slowly, to a total of 13.3 million around 1990, compared with 19.1 in the USA.

Northern and southern cultures

The nature and distribution of these retailing jobs in Europe reflects a recognised divide between countries with high and low densities of enterprise. To take an extreme example, households in Portugal consume less than half the goods of the average German household, but it has well over twice the number of retail enterprises per 10,000 inhabitants of Germany (Table 8.2). In the 'Mediterranean group'(1), Greece, Portugal and Italy have a very high density of enterprises, with more than 160 enterprises per 10,000 inhabitants. But along with Belgium (for historical reasons) and Spain, they average less than three employees per enterprise. This contrast can be attributed to the family-based structure of the enterprises (indicated on the table by the low percentage of employees), laws that protect small retailers, and the importance of tourism. Countries in the 'northern group' (2), such as the

158

Table 8.2 Employment characteristics in EU retailing, 1987–91

	Persons per enterprise	*Employees (%)*	*Female (%)*	*Part-time (%)*	*Hypermarket density*[a]
1 Countries with a low share of employees and female employment					
Greece	1.9	28.9	43.6	3.3	0.2
Italy	2.6	38.7	—	—	0.2
Portugal	2.1	46.8	45.6	5.3	0.2
Spain	3.2	46.6	53.7	6.4	0.3
Belgium	2.1	52.4	52.8	17.4	1.0
2 Countries with a high share of employees and female employment					
Ireland	4.5	70.2	45.6	13.4	0.2
France	4.5	74.1	50.3	25.3	1.5
Netherlands	6.7	79.1	56.7	47.3	0.3
Luxembourg	5.1	80.6	51.6	11.5	0.6
West Germany	4.2	84.0	52.8	38.3	0.9
United Kingdom	8.7	84.2	58.1	40.7	1.3
Denmark	5.4	84.6	64.1	37.3	1.3
E12	4.0	66.0	57.1[b]	24.2[b]	—
3 New members, 1995					
Austria	6.4	83.6	64.4	13.0	—
Finland	4.6	82.7	64.4	—	—
Sweden	6.2	82.8	60.5	37.3	—

Note: a. Hypermarkets greater than 2,320 sq. metres (1991) per 100,000 inhabitants
  b. E11
Source: CEC, Eurostat, 1993b, 1995a

UK, Germany and The Netherlands, have a highly concentrated retail sector with enterprises employing more than five persons on average. The proportions of both food retailers and sole proprietors are much higher in the Mediterranean, whereas multiple outlet chains, usually operated by food retailers, have penetrated further into the market in Northern Europe.

The form of employment, i.e. the occupational structure of retailing, depends on the form of ownership, principally on the 'stage' reached in *the shift from the southern to the northern pattern*, from sole proprietors to wage earners. *Department stores* offer a wider range of occupations to their employees. In addition to sales staff, which still have the dominant share, there is a variety of stock-room clerks, clerical, supervisory and managerial staff. *Supermarkets* and *hypermarkets* have quite different staffing patterns. According to French estimates (Kruse *et al.*, 1993), they comprise 50 per cent stockroom and check-out staff, 15 per cent managerial and supervisory personnel and 7 per cent food-processing workers. We must not forget *mail-order* retailing, which essentially employs clerical staff in large concentrations which have been important in providing work in 'assisted areas', for instance the UK sub-regions of Liverpool or Sunderland. (This form of retailing is not well developed in Mediterranean Europe.)

The generally accepted international model is one that sees the emergence in turn of department stores, chain stores, supermarkets and hypermarkets, and a 'fourth wave' is envisaged – see below. Each type of outlet can be compared to a product with a specific life cycle. The development of traditional department stores is stagnating, despite – or because of – their locations in city centres, and the ordinary supermarket has also reached a stage of maturity in all countries. Now the hypermarket has emerged, to threaten the smaller retailers with profound change in Southern Europe, as illustrated in recent dramatic change in Greece (Bennison and Boutsouki, 1995), but its growth has slowed down in France and Germany despite the energetic opening of new hypermarkets in the former East Germany. The density of hypermarkets per 100,000 inhabitants (Table 8.2) showed an expansion in E12 from 102 in 1975 to 733 in 1991, providing between 1.0 and 1.5 per 100,000 people in the leading countries, France, Germany and the UK, though large retail groups are now targeting markets in Mediterranean Europe.

The process of modernisation implies greater efficiency and higher activity. A trend towards the greater concentration of retail ownership, through acquisition activity and control of new sites, is reported in most countries, indeed 'the trend towards greater concentration is irreversible' (CEC, Eurostat, 1993b: 22). It is more marked in the food than the non-food sectors, and, as we would expect, more in the northern group of countries than the Mediterranean. The leading twelve groups have their headquarters in Germany, France or the UK, and they operate mainly in the food sector (Table 8.3). In my view, however, the increases of consumption and in labour costs are inducing the same *directions* of change in different countries.

The French example

France will illustrate the effect of the changing composition of retailing on that of employment. After very little change to 1960 (Baret and Bertrand, 1992), the country was one of the first to introduce large-scale distribution. By 1989, hypermarkets and supermarkets, led by Intermarche (with 50,000 employees), Leclerc (with 51,500) and Carrefour (with 76,200), had already captured a market share of over 50 per cent of food product sales. However, to face competition from large-scale retailers, small businesses organised themselves into franchise chains or buying groups. French customers remained fairly loyal to traditional neighbourhood stores and street markets, and the number of retail enterprises remained stable at 400,000 throughout the 1980s. The net result was that employment increased in crude terms from 1,595,000 in 1980 to 1,650,000 in 1990, but with a marked shift from self-employment to part-time and women's work. These last represented 30

Table 8.3 The leading EU retail groupings, E15, 1991–2

Rank	Group	Country	Main activity	Turnover (bn ECU), 1991/92
1	Tengelmann	Germany	Food	22.8
2	Metro	Germany	Food, mixed retailing	22.6
3	Rewe	Germany	Food	18.5
4	Carrefour	France	Food	16.2
5	Intermarche	France	Food	15.4
6	Leclerc	France	Food	15.3
7	Aldi	Germany	Food	13.4
8	Edeka	Germany	Food	13.2
9	Sainsbury	UK	Food	13.1
10	Promodes	France	Food	11.0
11	Tesco	UK	Food	10.8
12	Pinault-Printemps	France	Department stores	10.7

Source: CEC, Eurostat, 1995a

per cent and 75 per cent of employees respectively by the end of the period, concentrated in the food hypermarkets. A major problem was the gap between the large mass of lower skilled jobs on the one hand and the supervisory jobs on the other, giving little opportunity for career promotion (Baret and Bertrand, 1992). The effect of all these changes was that, in 1994, France had the smallest retail labour force in the EU, measured as a percentage of total employment (Figure 8.1).

EMPLOYMENT CHANGE

Change in France, with a small increase in total employees, but a marked shift toward part-time work (with its attendant problems), is not far removed from the European average. In the 1980s, there was a slow overall growth of retail employment in most countries, although Spain and Denmark registered reductions and Italy recorded an increase of 21.3 per cent even in 'full-time equivalents'. It seems that the modernisation of food retailing has made some inroads into productivity in most countries, offset by increases in consumption and the survival of smaller shops with lower productivity in other sectors. The combination of economies of scale, linked with the development of self-service, new information technologies and the rationalisation of structures and work processes, inevitably leads to higher productivity. It has negative effects on employment levels (Sparks, 1992) and 'the same could be said of the occupational and skill structure of the workforce from a qualitative point of view . . . a reduction in the proportion of sales personnel and a polarization of skills is likely to occur' (Kruse *et al.*, 1993: 19).

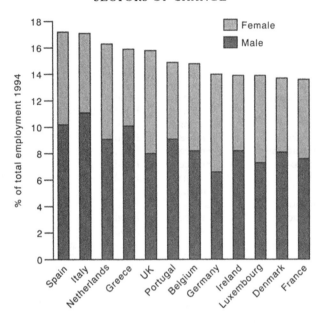

Figure 8.1 Dependence on employment in the distributive trades; retail, wholesale
and repairs, E12, 1984
Source: CEC, Eurostat, 1996a

The impact of supermarket and hypermarket conditions of employment is important and can be gauged from Table 8.2. In E12 by the turn of the decade, two-thirds of people at work in retailing were waged or salaried, but the self-employed and family workers still represented over half the retail workforce in Southern Europe, which had lower proportions of women in employment and very little use of part-time employees as such. Women make up more than half the retail workpeople in the second, northern group, although the use of part-timers remains distinctly different between countries, because of the varying cultural reasons and overhead costs that we mentioned in Chapter 6. The data

> suggests that part-time work is more widespread in countries where the modernization process is most advanced, but also that there seems to be a peak beyond which part-time work stabilizes or even decreases (Denmark, UK; the same can be seen in the USA). In Spain, the amount of part-time work is slowly increasing, but it is not yet widespread and is common only among large hypermarkets.
>
> (Kruse *et al.*, 1993: 26)

Multiple chains often prefer part-time workers in order to adjust staff levels to peak trading times and to control labour costs. For example, more than two-thirds of the 20,800 employees of the Swedish food retailing firm ICA were part-time workers (CEC, Eurostat, 1995a).

The companies' demand for flexible forms of work extends also to the employment of a young, mobile population, flexible contracts and the high proportion of women. Traditionally, many young people have always worked in retailing, and the youthfulness of staff may be correlated with a high rate of mobility. In France, for instance, retail is, like other sectors, receiving large numbers of young people, often, as we have seen from the unemployment data, for lack of better opportunities, and who do not stay long. Temporary employment is used for specific periods (evenings, vacations, Christmas) that may coincide with students' wish to work part-time. But several countries report a more general expansion of short-term contracts, sometimes with little or no social insurance cover.

The proportion of women as a whole has increased only slowly, while it is now stable in The Netherlands and UK, or even slightly decreasing as in Denmark. The proportion of women is much higher in sales and clerical jobs than in management (except in Denmark). This imbalance can be explained in different ways including mobility affected by the husband's higher paid job, and women's expectations linked with their socialisation.

Using all these dimensions of employment, competitive European retail employers have succeeded in holding wages below national averages. Given also the difficult working hours and conditions, there is a shortage of people willing to be trained for a long-term career in retailing, indeed the polarised structure hardly allows any promotion prospects.

Thus, in most of the countries for which data is available, the level of education and training in the retail sector is below the average. Even in Germany, where a great majority of young people go through one of the two apprenticeship systems, the profession of shop assistant is practised without relevant qualifications. Across the EU, the impact of regular, national training systems on retailing is very limited. Comparatively few recruits come from vocational schools, and much of the initial training is on the job.

It remains difficult to judge whether the enlarged market of 370 million people will engender the 'European consumer' with the same lifestyle and buying habits throughout the EU. Nonetheless, the sector is developing its own TNCs. In the late 1980s, retail groups in E12 started intensifying their international activities to prepare for this market (Treadgold and Davies, 1988; Treadgold, 1990), including European entry to Britain (Duke, 1993), and UK retailers 'increasingly recognising in Europe what they had hoped to find in North America' (Alexander, 1995: 80). French hypermarket operators exported their proven formula to the rest of Mediterranean Europe. The highest international sales were made by the German food groups, Tengelmann and Metro (Table 8.3). Internationalisation is limited, however, to a few retail groups and countries (Robinson and Clarke-Hill, 1990), and

Sainsbury and Marks and Spencer sold only about 12 per cent of their goods outside the UK. Further expansion will diffuse the latest methods of controlling staff more widely, but meanwhile Britain stands as the most extreme example of the concentration of ownership in the EU, and the most dependent on part-time employees.

THE EXAMPLE OF GREAT BRITAIN

Despite a one-off expansion in the 'Lawson-Thatcher boom years' of 1987–9, retail payrolls in Great Britain stood at the end of 1995 no more than 12.1 per cent higher than twenty years earlier. This rate of increase was smaller than in other countries and smaller than in service activities. Despite the job creation claims of most new superstores, retailing is not a net provider of many new jobs.

Yet this is just what is of special interest. Severe competition has forced up productivity and generated new levels of numerical flexibility in retail employment. New patterns of capital investment have concentrated spending and jobs at further remove from housing. The government finally set the brakes on 'out-of-town' centres in a policy decision of 1994.

Investment and productivity

The trend towards concentration had been present for the previous twenty years, but the 1980s were characterized, most notably, by the emergence of a small group of retail corporations whose turnover, employment levels, profitability and sheer market and political power came to rival the largest industrial corporations in any sector of the UK economy.

(Wrigley, 1993: 41)

Changing market conditions encouraged individual companies to expand through both acquisition activity and new building. Expansion was heavily dependent on the climate for institutional investment of 1983–9, the years between the two massive recessions of the Thatcher-Major period. Within the property investment sector, retail development acquired a particular attraction, peaking in the years 1986–9 which saw a 60 per cent increase in the total return on retail investments.

Despite high levels of capital investment, labour costs contribute up to 50 per cent of retailers' gross margins. The transition from smaller, labour-intensive, to larger, capital-intensive forms of outlet was reflected in an intensification in the use of labour. Indeed, we know in general that retailing has emerged as one of the centres of productivity growth in the British economy. Wrigley (1993) showed that overall labour productivity (as measured for 'full-time equivalent' staff) had increased by 21.6 per cent from 1961 to 1971 and 29.7 per cent from 1971 to 1982. For the period 1983 to 1987 he reported an increase of 12 per cent, accompanied by an accelerating use of electronic equipment at check-outs.

Changes in labour structure – flexibilisation

The main ingredient of these improvements in productivity has been part-time employees, for example:

> In common with other retailing companies, Sears has been trying to cut its costs by doing away with large numbers of full-time posts and using part-time staff to provide cover at busy periods. By the end of 1991, around 60 per cent of Sears' employees were already working less than a full week. Now the proportion is 65 per cent, said chief executive Liam Strong. Burton has embarked upon a similar strategy. At the beginning of this year, the group said it would be axing around 1000 full-time jobs and replacing them with around 3000 part-time posts. In 1988, around 60 per cent of Burton employees worked full-time; by 1992, the number of part-timers and full-timers were roughly equal.
>
> (B. Laurance, *The Guardian*, 20 June 1993: 17)

Evidence for eleven anonymous retail companies, and three in wholesaling, was provided by the Distributive Trades Economic Department Committee (1988). In retailing companies, the proportion of part-time employees varied between a quarter and three-quarters. In the four largest corporations, employing between 15,000 and 55,000, the proportion was over two-thirds. This significant difference is closely associated with the charge of de-skilling, which is in fact much more relevant here than elsewhere: self-service retailing has reduced the specialisation required previously in a shop assistant's job and has reduced the bulk of the tasks within larger stores to shelf-filling or till operation (Sparks, 1983).

The main need, in the context of intensive competition and squeezed margins, particularly in food multiples, was to match manning levels over the day and during the week to fluctuations in consumer demand. The greater use of part-timers started as a convenient way of filling peak times during day-time pressures, but, with the additional spur of longer opening hours in the evening that would otherwise entail overtime bills, they became an essential norm to the running of a large shop, for instance in three shifts. Thus, retailing was one of the activities studied in the seminal paper of Atkinson and Meager (1986) on flexibility.

Although paid at the same basic hourly rate, part-timers' total hourly earnings are commonly less than those of their full-time colleagues. Part-time women in retailing earned 11 per cent less than the hourly earnings of women full-timers and 40 per cent less than men (Distributive Trades Economic Development Committee, 1988). The same study goes on to defend retailing against criticisms of companies' savings from part-timers' exclusion from pension schemes, sick pay, maternity and other benefits, or the companies' aim of avoiding threshold levels that trigger National Insurance contributions and employment protection rights. However, these must be regarded as basic factors that underpin Britain's greater use of part-timers than all other EU countries than Denmark. Even after adjusting for this factor, British companies fall behind European ones to a surprising degree when measuring employee productivity (Economist Intelligence Unit, 1993).

Employment trends over time

Basically the growth of productivity has absorbed the effect of increased consumption on shops, and prevented retailing becoming a major area of employment growth. Including part-time workers in the crude total of employees, Figure 8.2 demonstrates remarkable stability in overall employment levels, with a cyclical response that is surprisingly moderate. However, it shows a trend of moderate expansion if measured trough-to-trough from 1983 to 1993.

Part-time employment grew faster, increasing its share of retailing employees from 35.0 per cent in 1971 to 55.5 per cent by end-1995, including a little-known feature, the growth of men's part-time employment in retailing, at a proportionately faster rate than women's. At the same time, women's

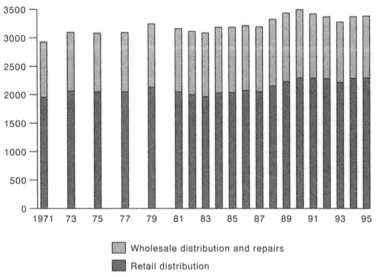

■ Wholesale distribution and repairs

■ Retail distribution

Figure 8.2 Time series of employment in the distributive trades, thousands, GB, June 1971–95
Source: Employment Gazette

full-time employment was sustained until the years 1987–9, but fell back decisively in the recession after 1990. This latter period involves not just the substitution of part-timers for full-timers, but also direct re-designation of workers from full-time to part-time, leaving two-thirds of all women retail employees as part-timers. These features were common to women's employment in the economy at large in this last recession, from which women's part-time employment totals emerged relatively unscathed.

These divergent trends in working conditions and hours are creating a more distinct divide between the relatively small numbers of executives and managers and the larger number of less skilled workers. This has implications for workers' conditions, prospects and social life, but the ratio of full-time and part-time jobs will vary between different parts of retailing and geographical areas of the country. The way in which services may differently exploit the labour force of different areas is central to a view of 'the social construction of space'.

Sectoral and geographical aspects

Disaggregation is essential. The principal picture arising from Table 8.4 is one of re-distribution between sub-sectors. Mixed businesses, confectioners and off-licences continued 'the decline of the corner shop', while food retailing demonstrated the largest increase in numbers. A similar contrast is evident in every region. Previous analyses (Reynolds, 1983) have found a general relationship between changes in retail employment and that of population,

Table 8.4 Trends in retail sectors, GB; employees by 'activity heading', 1981–94 (thousands)

	Sept. 1981	Sept. 1991	Change 1981–91		Sept. 1994
			Nos.	%	
Food retailing	564.0	720.1	+156.1	+27.7	702.9
Confectioners, off-licences	158.7	110.2	−48.5	−30.6	119.5
Chemists	125.0	125.9	+0.9	+0.7	125.8
Clothing	148.9	175.7	+26.8	+18.0 ⎱	221.1
Leather goods	58.6	68.3	+9.7	+16.6 ⎰	
Household textiles	22.1	30.9	+8.8	+39.8 ⎱	247.2
Household goods	175.3	239.6	+64.3	+36.7 ⎰	
Books, stationery	68.4	70.7	+2.4	+3.4 ⎱	
Other retail	106.5	204.7	+98.2	+92.2	583.5
Mixed	350.2	292.5	−57.7	−16.5 ⎰	
Motor trade	192.6	211.2	+18.6	+9.7 ⎱	279.7
Filling stations	79.0	61.3	−17.7	−22.4 ⎰	
Total, retail	2,049.3	2,311.1	+262.0	+12.8	2,280.7

Source: National Online Manpower Information System, *Employment Gazette*

167

though with departures between 1971 and 1981. These were attributed possibly to delays in the restructuring of retailing in certain areas, or to the role of retired people (with lower spending needs per head) in the growth of the South West.

The overall regional pattern for the years 1981–91 (Figure 8.3) shows two regions with significant increases of population, the South West and East Anglia, to have high rates of increases in retail employment, although Wales also expanded by 20 per cent. Otherwise we can say that none of the six Regions of population decline saw an increase of more than 15 per cent in numbers of total jobs, or of 30 per cent in part-time jobs. This reaffirms a relationship established in Townsend (1986) between the distribution of increase in part-time jobs in general and that of the resident population across the regions of Britain as a whole. Part-time employment is slightly more common in the most rural areas, as part of a finer-grained and more systematic gradation between urban and rural areas as a whole ('Townsend *et al.*, 1995). This has already been observed by Reynolds (1983: 357), whose

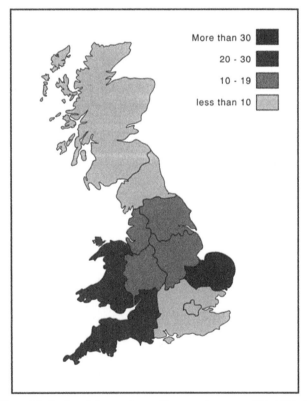

Figure 8.3 Regional change in retailing employment, GB, 1981–94
Source: *Employment Gazette*

data 'reinforce the impression that the movement of population from the major conurbations and into outer "rings" is associated with similar movements in food retailing employment, and particularly of part-time employment'.

Significance of the trends

Distribution, as a service activity, is now reflecting and reinforcing the re-distribution of the wealthier British population. Change in customers' time-space diaries, due itself largely to rising women's activity rates, together with further increased car ownership, have transformed shop opening hours and strengthened the need for a workforce that is half 'part time'.

Many strains and pressures remain to be absorbed from these changes. Before its abolition, the Distributive Trades Economic Development Committee of the National Economic Development Office (1988) was so concerned about some aspects of part-time employment as to mount a study at establishment level. It found part-time recruitment to be sensitive to local labour market conditions, for instance in using student workers for certain areas and times. It identified major issues over part-timers' training and possible promotion. Beyond the necessary minimum, training was not available to part-time workers, partly because of the problems of arranging it for different shifts. This fed in turn into the debate about promotion. While many part-time workers claim to be well satisfied for the time being, many, equally, are working well below their educational level. Their scope for promotion is severely restricted or non-existent because of the divisive effect of their differing hours on the teams they would be required to lead.

The evidence shows some differences between sizes of firm, with longer service and better benefits being prevalent in small firms. Penn and Wirth (1993) emphasise from a case study of the Lancaster area how many variations there are even between five main stores in one area. All firms saw disadvantages in employing part-timers, in addition to the advantage of flexibility. Yet the same set of perceptions co-existed with radically different management policies, over attitudes for instance to married women workers, in leading firms.

> Indeed, on the basis of these data we conclude that the precise social effects of the growth of female part-time employment in retailing during the 1980s were more complex than most commentators have suggested and that managerial recruitment systems are far more company-specific than many had imagined.
>
> (Penn and Wirth, 1993: 265)

Notwithstanding its acute exploitation of women and young people as shopworkers, the UK appears to be only the extreme example of a fairly uniform direction of change in Europe.

FUTURE TRENDS

Notwithstanding the power of computing applications and automatic reordering of stocks, which has itself contained the growth of employment, it is argued that 'retailing remains a people business. Its most successful practitioners must combine computing and logistical prowess with merchandising skill, customer service and staff empowerment' (*The Economist*, 4 March 1995: 23). A return to higher levels of customer service is one subject of speculation for the future. In particular, Walsh (1992) argued that the impact of the Single European Market programme might be to secure a general upgrading of the quality of retail services. More sophisticated consumption patterns, greater disposable income and more leisure time might all lead to a demand for higher quality services, with all-important consequences for the employment, skills and labour utilisation of employees, with more full-time women workers.

Kruse *et al.*(1993) also argue that the trends towards de-skilling in the occupational and skill structure of retailing may be reversed in future, in the *'fourth wave'* of retailing change. Such authors identify future change with a wide variety of developments, including a greater diversification of types of store, with high volume sales of discounted goods at one extreme to 'niche selling' of specialised goods such as ties, or the provision of more service in supermarkets at the other. Other authorities, such as Fernie (1995), place most emphasis on a 'fourth wave' comprising warehouse clubs (such as developments of Costco in the UK), factory shopping malls and airport retailing complexes, with teleshopping, or on-line shopping from home, still an uncertainty.

However, some European analysts predicted only a modest growth of consumer spending for 1995 to 2000, with higher indirect taxes, fluctuating interest rates and continued anxiety about job security creating a generation of cautious customers reluctant to part with their money. Added to that are further effects on employment resulting from the expected growth of teleshopping, new management of information systems and the ultimate development of automatic checking out, whereby the customer generates his/her bill by passing through an area where the bar codes of purchases are read at a distance by new kinds of equipment.

On the other hand, there would appear to be less radical change ahead in shop location and the location of stores. The success of out-of-town regional shopping centres and the damage to town centres are very clear in England (Howard, 1992; Lowe 1991), while not all of the US shopping centre formats have proved successful in Europe (Thomas and Bromley, 1993; Davies, 1995). Administrative restraint by government land-use planning authorities has followed in both cases, although outstanding planning permissions extant at the time of the UK government's change of view are a major factor in allowing more out-of-town centres.

SUMMARY

- Leisure and consumption are very significant for employment levels and practices.
- Differences between EU countries in the pattern and intensity of retailing, notably between northern and Mediterranean Europe, are gradually disappearing as the French or UK type of hypermarket takes an increasing share of trade.
- Retailing is a leading edge of change in labour practices and social relations, being ahead of other activities in flexibilisation.
- The part-time employment of women, with all its problems, has probably passed its peak in the USA and may be nearing it in northern Europe.
- A great deal may be about to happen. Numbers of new out-of-town hypermarkets may falter, and a 'fourth wave' of change may break in a welter of niche marketing, factory shops and tele-shopping.
- Changes in the UK distribution of retail jobs reflect and reinforce changes in the map of population toward rural areas.

9

A TOURIST SOLUTION?

Why does tourism sit at the forefront of thinking in regional development? Above all, because other activities cannot provide the jobs, especially for the young (Pearce, 1988). 'The growth of large-scale unemployment in many of the industrialised European countries has led to governments assessing tourism in a new light' (Shaw and Williams, 1990: 240).

I shall trace the emergence of new forms of tourism, mainly away from the coast whether urban or rural, from their beginnings and ask what contribution tourism can make to particular areas. On the generous side, we can include visitors' spending on a variety of services. On the other, competition between new attractions must be borne in mind for visitors' spending. And what about quality of jobs, a serious problem in tourism as another prime example of flexibilisation? France and the UK are examples inspiring caution about urban tourism as a solution. And what are the constraints on growth from environmental limits and from competition overseas?

OVERALL SIGNIFICANCE

The 'production of tourism' involves the projection of cultural images by a sophisticated set of attitude-forming specialists as well as the building of facilities by all the professions involved in construction and in the packaging of holidays by agencies, operators and hoteliers. Consumption and leisure spending have increased so much in Europe that holiday travel has become a principal channel in redistributing income between countries and regions. As compared with shopping, tourism, travel and holidays involve much longer-distance movement (Jansen-Verbeke, 1990), but have an even stronger tendency for employment to consist of low-paid, part-time, seasonal and, in hotels and catering, women's jobs.

The EU has frequently expressed support for the transfer of income involved in the annual migrations of holidaymakers from wealthier countries of northern Europe to the poorest regions of the Mediterranean (Cole and Cole, 1993). Individual countries, long concerned with their financial balance of payments from tourism, have tried to maximise tourist activity from the viewpoint of the overall economy. Individual regions have incorporated tourism in strategic physical plans and invested in promotional agencies. Last but not least, former industrial areas suffering from the deindustrialisation of their economic base have adopted tourism, including the exhibition of the past industries themselves in their heritage museums, on the argument that tourism is a growing activity. Where others are declining, at least this sector allows for more people to make a living in Europe.

GROWTH FROM SMALL BEGINNINGS

Tourism was from the start based on wealth and cultural constructions. The habit of taking a pilgrimage, to a centre such as Delphi, Rome, Santiago de Compostella (in Galicia in north-western Spain) or Canterbury, is generally credited as having contained a touristic element. It certainly provided a modicum of employment in the service classes of the day at the destination (some in clerical garb), and in providing lodging and transport en route. Later, the 'Grand Tour' of European cultural centres by English gentlemen provided an early kind of 'urban tourism', and the establishment of spas based on the consumption of spring mineral waters shows us how a whole town, designed in the leading styles of contemporary grand building, could attain a thriving economic base from the spending of visitors. The first modern example of these gave its name to the group, that of Spa (now in eastern Belgium), followed by numerous large examples such as Karlsbad (now in the Czech Republic), Baden Baden (Germany), Aix-les-Bains (France) or Leamington Spa (England).

Before the nineteenth century, sea bathing or mountain climbing inspired horror rather than attraction. The popularisation of sea bathing by the Prince Regent at Brighton was followed by the systematic promotion of coastal resorts by the railway companies of an industrialised Britain, using both weekly and day return tickets; no writer, of economic geography texts disputes that these resorts had a viable economic base among the specialised towns of Victorian Britain. With the diffusion of railway development across Europe, other countries followed the British example, with resorts on the Belgian coast or the French/Italian Riviera. Meanwhile, the British upper classes had popularised Alpine climbing and ski-ing, and lake and mountain resorts in Switzerland and Austria added to the phenomenon of inland tourism, along with the growth of some walking, cycling and motoring holidays between the wars.

However, before package air holidays and the near-universal extension across the social scale of holidays with pay, international tourism in Europe was comparatively small. As recently as 1960 no European country received more than 10 million foreign tourists, whereas by 1992 the leading destinations, France, Spain and Italy, respectively received 58.5, 36.1 and 27.0

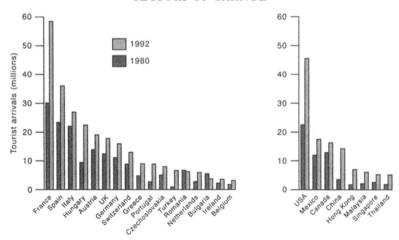

Figure 9.1 Growth of international tourism, 1980–92
Source: World Tourism Organisation, 1993

million tourists, within an E12 total of 188.4 and an all-European total of 287.5 million (Figure 9.1).

Europe is the world's leading centre of tourism, for this total in turn represents 59.6 per cent of the estimated total of 475.6 million world tourist arrivals, itself expanded by 70 per cent since 1980 (World Tourism Organization, 1993). But the 59.6 per cent is less meaningful than many think, for a continent with many small countries will automatically have more 'international' tourism than will large federal states, notably the USA, where many long-distance movements (including those to Puerto Rico, Alaska and Hawaii) will be counted as internal.

The demand for holidays provides the most fundamental approach to understanding these large increases in industrialised countries. Socio-economic influences on access to tourism and leisure are shown (CEC, 1987) by strong correlations between growth of GDP per capita and the proportion of people taking holidays. Furthermore, tourism is characterised by a 'positive income elasticity of demand', that is to say that the demand for holidays has risen in the West proportionately more than has personal income. At any one time, the propensity to take foreign holidays is markedly affected by social class and age. And while the total number of holidays of four nights or more taken by EU residents is estimated by CEC (1993b) at 250 to 300 million for 1991/2, in France and the northern EU countries more than 60 per cent of the population take holidays, while the levels for Mediterranean countries' residents stand significantly lower.

Among the principal geographical patterns (Burton, 1994) has been a change in which air tourism has helped to deepen and broaden the northern Europeans' experience of Mediterranean areas, with movement extending

174

further south over the Iberian mainland to Andalucia and southern Portugal, and mass air travel expanding tourism in Majorca, Sicily and Greece, as well as Madeira, the Canaries, Morocco, Tunisia, Cyprus and Turkey. Over the 1980–92 period, Figure 9.1 shows the fastest growing destinations in percentage terms to have been Turkey, Portugal and Hungary, although in absolute terms, the biggest increases of tourist arrivals were attracted by France and Spain, which remained at the head of the European league table. In the late 1980s, there was a most interesting expansion in the number of tourists from southern to northern Europe, but in the early 1990s flows were fairly static within the EU, though expanding between eastern and western Europe. The CEC (1993b) estimated the plane to be involved in 35 per cent of European international trips (leisure and business) in 1990.

Much of the rise of air travel in the last two decades has been associated with the increased use of wide-bodied jet planes. While opening up more of the Mediterranean coast and islands, the trend has also taken Europeans outside Europe on an increasing proportion of long-haul holidays, and brought more visitors from other continents. The outcome of this wider movement is actually unfavourable for the EU, as its balance of trade with the extra-EU world has been shrinking since 1985. The EU balance with the USA became negative by 1992 while the net deficit with the other 'industrialised' countries increased (despite a boom in Japanese visitors), and a deficit with low income countries appeared (CEC, Eurostat, 1995c). Overall, Europe's share of world tourist arrivals has declined from 72.5 per cent in 1960 to 59.6 in 1993, causing great concern in some quarters but not surprising in the light of economic growth and tourist success in, for instance, Hong Kong, Malaysia, Singapore and Thailand (Figure 9.1). The increasing number of new Asian resorts being built on the basis of a regional economic boom may constitute a longer term global threat, especially seeing that 'thanks to local labour, guests in resorts outside Europe are pampered by battalions of informed staff, a heady change from a solitary, surly chambermaid' (*The Economist*, 29 July 1995: 5).

DIVERSIFICATION INTO NEW FORMS OF INLAND TOURISM

More temperate locations have benefitted from the movement away from mass tourism products toward more specialist and niche holidays, incorporating the increasing popularity of split and short break holidays and development of cultural and rural tourism . . . in the 1990s the direction of tourism flows and expenditure within the EC are more uncertain than they have been for some time.

(CEC, 1993b: 19–20).

These directions of change are allowing more localities of the EU to make a living from tourism, possibly in a greater variety of employment and perhaps increasing the ratio of domestic to international visits within respective countries. What then is the relative strength of traditional coastal locations? What factors prompt new kinds of inland and urban tourism?

'In Europe, taking holidays at home is still more popular than travelling abroad, even though international tourism seems to grow faster than domestic tourism' (World Tourism Organization, 1993: 3). Of European adults taking a holiday in 1992, almost 60 per cent took this in their own country. Figure 9.2 shows how more than 50 per cent of nights in the hotels of Germany, Sweden, Finland, Norway, Italy and France were spent by domestic visitors. Second holidays and short breaks of one to three nights are increasing at a much faster rate than main holiday trips. This relates partly to the growth of disposable income and partly to innovative packages to fill off-peak capacity and to compensate for unexpected (but usually short-lived) downturns. There has been a lowering in the average length of stay and a separating out of distinctive market segments, including the over 55s, eco-tourists and seekers after fitness or health.

Together with a faster growth of winter than summer travel, these changes mean that 'the traditional summer holiday in a Mediterranean destination has reached the mature phase of its growth, indeed may even be into

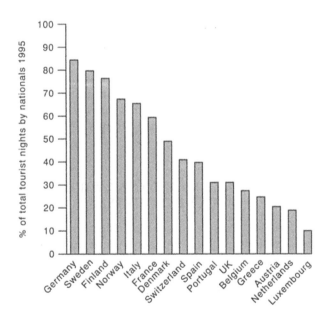

Figure 9.2 Dependence on domestic tourists in western Europe, 1993
Souce: CEC, Eurostat, 1995b

a declining product phase' (CEC, 1993b: 18). Likewise, 'the most exposed part of the industry could be Mediterranean Europe's sun-and-sea resorts', (*The Economist*, 29 July 1995: 56), suffering saturation of facilities, the environment and the workforce at a time of increasing skin cancer.

In Britain, France and Spain alike the interest is in making the most of other kinds of attractions, whether old-established museums of high culture or new activities in inland centres. Farm tourism (mainly bed and breakfast) has become an important activity of farmers' wives, gaining new signifi-cance from downward pressure on farm incomes under the Common Agri-cultural Policy, alongside policies for the temporary 'setaside' of fields, and changes in support for upland farmers (Dernoi, 1983). Grants have vari-ously been available for diversification into leisure activities, including golf ranges and accommodation (as well as forestry and small business). Other developments in rural settings have included the welcoming of paying visitors to historic properties, second homes, wildlife parks, leisure and coun-try parks and gardens, as well as new activities such as the development of dry ski slopes.

The re-marketing of historic cities has become prominent (Ashworth and Voogd, 1988) as in The Netherlands (Ashworth and Bergsma, 1987), but the most striking urban feature has been the use of North American precedent in employing tourism to provide city regeneration and employment (Law, 1994). Both in the USA and in Europe, large cities have always been import-ant centres of culture and entertainment, but largely unrecognised as such. However, if we use present-day definitions to include business visits and conferences, city tourism was extensive even before short-break holidays were asked to improve a city's jobs, environment, image and civic pride. The latest models were provided by the waterfront redevelopment of Baltimore and Boston in the USA, including marinas, aquaria, conference pavilions, boutiques and entertainment. They may not have done much for surround-ing housing areas or for the quality of jobs, but they have certainly attracted millions of visitors, produced some employment and been imitated from Barcelona to Hartlepool.

Thus tourism has become seen as a tool of European regional develop-ment in new ways. In addition to the inclusion of business tourism through conference facilities and hotels, the attraction of day visitors from one region to another has become a significant feature. In the US the spending effect of college football matches has made them into tools of area development policies; but sport is just one of dozens of means of attracting day visitors from one area to another. In Great Britain, the Leisure Day Visits Survey (Baty and Richards, 1991) established a figure of 379 million journeys for pleasure over 40 miles in 1988/9, compared with a figure of 34.5 million tourists going abroad and 17.9 million arrivals in 1992. In any case, it is generally impossible to distinguish employment from leisure, day tourism and overnight tourism in available statistics, not least because all three kinds of activity may entail people drinking in the same bar together or visiting the same museum.

EMPLOYMENT IN TOURISM

Tourism is relatively labour intensive and generates more jobs than its share of national income might suggest; it is most prevalent in the richest countries themselves, and so it is central to this book. Taking a global estimate, the World Travel and Tourism Council estimated that tourism provided direct employment for 118 million people, or 6.5 per cent of world employment (CEDEFOP, 1994), and there are similar claims for it being the world's largest industry. I agree with Shaw and Williams (1994) that these claims are examples of a gross tendency to overestimate the size of tourism, in the face of many problems of definition and data.

Nonetheless, a study carried out for the European Parliament and cited by CEC, Eurostat (1995c: 2) credited the tourist sector with 5.5 per cent of GNP and 6 per cent of employment – 7.5 million people. However, the accompanying claim that it was thus the 'largest economic sector in the EU' depended on a comparison with individual manufacturing industries that were narrowly defined: the broad category 'engineering and metal manufacture' alone employs 14.2 million.

Most E12 countries had between 5 and 7 per cent of the total workforce in tourism in 1990, exceeded not surprisingly in Spain and Portugal (Figure 9.3; note the *wider* range in hotels and restaurants). Williams and Shaw (1991) report a similar range except that the new, small EU members show more extreme values, with Austria and Switzerland respectively credited with 18 and 14 per cent of their total employment in tourism, but the Scandinavian

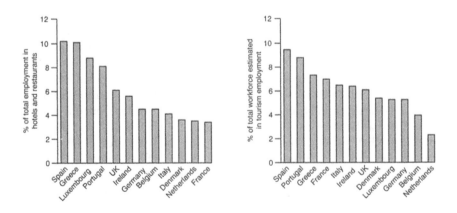

Figure 9.3 Dependence on hotels, restaurants and tourism employment, E12
Source: CEC, Eurostat, 1995b

countries with only about 2 per cent. Beyond these broad estimates, which form the currency of politicians' discourse in the field, great care is needed in trying to match available national data with the objective in hand, and my next sub-sections show a range of qualifications that build up the necessary critical view.

Composition of employment in tourism

Employment varies greatly according to the type of tourism, especially bearing in mind the element of self-catering in camping, caravanning and time-share accommodation. My ideal would be to identify the direct employment of tourism as comprising not only jobs resulting from overnight tourists visiting an area for pleasure, but also from business and conference trade *and* from day visits of more than a local kind. In practice, the readily available data give either too small or too large an estimate of the volume of tourist employment in a given area. For instance, in Belgium the total employment in 'accommodation establishments and other tourist activities', at 132,500, is about ten times the size of the component for 'hotels and similar establishments' (CEC, Eurostat, 1991). The Eurostat definition also incorporates travel agencies, car hire, tourist offices, libraries and museums and all bars, restaurants and coffee houses.

Similarly, Great Britain in mid-1995 had 344,000 employees in hotels, campsites and short-stay accommodation, but 1,548,000 in 'tourism-related activities' (*Labour Market Trends*, Table 1.14, quarterly; the figure had dropped seasonally to 1,479,100 by end-year). This definition includes sport and other recreational activities, but excludes tourist spending in shops, theatres and cinemas. In including, for instance, all public house employment, it will not stand the emphasis placed on it by politicians and the industry (Hudson and Townsend, 1992).

It appears eminently more reasonable, if seeking a single national figure, to abstract different proportions of all the possible sectors, as determined by tourism's share of their overall activity. For example, Medlik's (1988) formulation incorporates 42 per cent of hotel and catering jobs, 25 per cent of recreation employment, and smaller proportions of, say, retailing and transport, to produce an estimate for Great Britain reduced in net terms by 30 per cent to just over 1 million jobs, 5.1 per cent of the total (compared with 18.0 per cent for manufacturing).

The flexible component of tourism jobs

The bias of tourism jobs toward seasonal, part-time and informal flexibility deserves consideration in its own right and further affects our estimates of the importance of tourism. Although the development of transnational and European chains is prominent, there are many situations where owner-run pensions and farm bed-and-breakfast are far more important, not least in Italy as a whole (King, 1991). Referring to the EU:

> it is clear that the great majority of tourist facilities are run by small and medium-sized businesses, which often tend to be fragmented into a system of micro-businesses concentrated in tourist areas but also scattered around the rest of the country. This particular morphology – which also has a marked influence on the structure of ownership (self-employment) and on the level of education and skills of those in the working sector – is to be found in all the countries surveyed, even if the pattern differs slightly.
>
> (CEDEFOP, 1994: 15)

This is an aspect that is important in Southern Europe, where, for example, self-employment accounts for 42.3 per cent of all Spanish tourist activities, compared with 29.9 per cent in Germany. It is proportionately less important in the UK but in this case the estimated numbers of self-employed would add 188,000 to the total of 1,084,000.

The correction of part-time jobs to full-time equivalents would, on the other hand, make for a major reduction. In the service sector as a whole in E12 more than a third of all women working in the service sector are employed on a part-time basis (CEC, Eurostat, 1996c), with average hours of all women in distributive trades and hotels standing at thirty-two per week. A full set of evidence for the 'tourism-related industries' is available in the British case (Table 9.1), where part-time workers are those defined as

Table 9.1 The incidence of women's and part-time work in tourism-related activities, GB, December 1995 (thousands)

	Part-time		Female		Total
	no.	*%*	*no.*	*%*	*no.*
Hotels, campsites and short-stay accommodation	142.9	45.7	185.2	59.2	312.9
Restaurants	186.0	57.3	177.8	54.8	324.6
Bars	279.8	72.3	249.3	64.4	387.1
Travel agencies and tour operators	18.2	24.7	54.4	73.9	73.6
Libraries, museums and cultural services	35.6	45.3	52.0	66.2	78.6
Sporting and recreational services	146.5	48.5	150.2	46.7	302.3
Total, tourism-related industries	809.0	54.5	868.9	58.7	1,479.1

Note: The above relates to employees in employment; estimates for additional self-employed workers indicate a total of 188,000.
Source: Labour Market Trends, 1996, 103, 11: S (also quarterly)

working less than thirty hours per week. In Britain, all the tourism-related activities except travel agents rely on part-time employees for more than 40 per cent of their staff, although there is a very marked distinction between hotels and restaurants, on the one hand, and bars, which are obviously more dependent on part-timers, on the other. In all no less than 66.0 per cent of women workers in tourism-related activities in mid-1995 worked part-time, and 54.5 per cent of the total.

If we convert to full-time equivalents by the method endorsed by Medlik (1988: 60; similar to Townsend, 1986), counting part-time employees as half units, then this amounts to a reduction of 27.3 per cent in the total of 1,084,000 (above), suggesting a workforce equivalent to (788,000 + 189,000 =) 977,000. This agrees with Medlik's method of calculation if we say that tourism has the equivalent of about 1 million jobs in Britain.

Qualitative aspects

This quantitative consideration must be married with all the qualitative concerns that exist about the conditions of tourism work, which embrace deep-seated cultural issues about providing consumer spectacle and consumption, now that they are driving forces of social life: 'service workers must increasingly conform to the highly prescribed, almost 'theatrical' requirements of working in fast food restaurants, or tourist and entertainment facilities (Marshall and Wood, 1995: 168).

Urry (1990: 70) argues that the sort of service required in providing the intangible ambience of a friendly hotel or restaurant constitutes 'emotional work', undertaken by relatively low level workers who are usually women; 'overlying the interaction, the "service", are often particular assumptions and notions of gender-specific forms of behaviour' (Urry, 1990: 71). Bagguley argues that:

> the hotel and catering sector has specific economic and social charac-
> teristics which have meant that 'flexibility', both functional and numer-
> ical, is much more highly developed than elsewhere in the economy,
> and furthermore, that this labour flexibility cannot be understood or
> explained without a systematic consideration of gender relations in the
> workforce.
>
> (Bagguley, 1990: 3)

Gender and flexibility

The majority of world tourism workers in the mid-1980s were shown to be women (Kinnaird and Hall, 1994). This was the case in nine of E12, with Scandinavia attaining proportions of up to 80 per cent women. In Greece, 'jobs in accommodation are dominated by women (bed-makers, cleaners, servants, etc.) especially in unrecorded rooms and the informal sector, where such jobs come "naturally" as an extension of housework' (Leontidou, 1994: 102), indeed the attraction of housewives into tourism has been government policy. In Ireland, Breathnach *et al.*(1994: 71) conclude that women's occupation of the part-time and unskilled jobs of this highly seasonal and unstable sector 'is primarily a reflection of the generally marginal status of women in the Irish workforce', while the employment and regional development effects leave a lot to be desired (Deegan and Dineen, 1992; Hannigan, 1994).

Shaw and Williams (1994) report strong occupational segregation in most societies, with women tending to be allocated the commercial tasks that are similar to household tasks (serving meals, working in kitchens and making beds). Managers use this social construction of women's work, and the possibility of keeping them on temporary contracts, as the means of structuring the internal labour market in tourism. Women, as cooks, barstaff and cleaners, provide tourism with numerical flexibility, while the men enjoy greater training and functional flexibility (Bagguley, 1990), and, through their strongly shared values and norms, effectively exclude women from managerial positions of power (Hicks, 1990).

More than in retailing, tourist demand varies by the hour, day, week and month, sometimes unexpectedly. Although some cultural barriers to flexible/insecure working were identified, Guerrier and Lockwood (1989) found abundant opportunities for employers to profit from insecure jobs. In Willington (Chapter 1), the daughters of miners who earned 'good money' wait on standby to serve or wash up in an expensive restaurant with a regional clientele. Bagguley (1990) emphasises that such flexibilisation dates back to the 1960s and originated in labour shortage! Kitchen work was significantly transformed by technical change, by convenience foods and new electrical appliances in the 1970s and by 'fast food' in the 1980s, which together reduced the overall numbers of women kitchen workers and brought in more workers under the age of 21. In Britain the early withdrawal of Wages Council protection for this age group emphasised the trend. Macdonald (1996) found a quarter of her sampled employees in central Liverpool hotels to be under 25, many unmarried and living at home still with parents. Urry (1990: 80) recognises, however, that there are staff who do not fit a division between a core and peripheral workforce, even though 'there is plenty of evidence to support the view that flexible working practices have for some time been a key feature of tourist-related industries in Britain'.

Especially in capital cities and in Switzerland (Gilg, 1991), hotels appear to make great use of transient staff, often recruited from abroad, while migrant workers to the resort coast are common in countries as diverse as Wales (Ball, 1989) and Portugal. They add to especially high turnover, and make it difficult to sustain adequate skill levels and training programmes. A relatively weak and un-unionised labour market leaves the workers exposed to low pay, with far fewer on top salaries than in other activities. Internationally comparable EU data are not available for services. In Great Britain *(Labour Market Trends,* Table 5.5, quarterly), hotels and catering suffer the lowest hourly and weekly average earnings for non-manual workers (£6.30 per hour in 1994) and for manual staff (£4.24 per hour, with women on an average of £3.81), considerably below the levels of the distributive sector. Planners must have very considerable reservations about the concept of employment development through tourism.

DYNAMICS OF CHANGE IN THE 1990s

Growth at low levels of productivity and earnings

'Many feel the arguments for demand-led growth in consumer service employment are less convincing than for business services' (Marshall and Wood, 1995: 160). In general this may be true, notably for retailing as found in Chapter 8, but tourism employment resumed its pattern of growth with great resilience in quantity if not quality after the recession of both the early 1980s and 1990s. Figure 1.5 of this book (page 21) displays recreational activities and hotels and restaurants among the only ten E15 activities in which employment grew from 1990 to 1994 (from a total of thirty-five activities, as defined by CEC, 1995b). Job growth in each was then at 2 per cent per annum, still a considerable fall compared with 1985 to 1990. Retailing and distribution, in contrast, were among the fourteen sectors whose employment declined after 1990. The only sector to expand consistently in both periods in the E15 (by 7 per cent per annum), was 'producer services' (Chapter 8). In Great Britain, however, 'tourism-related activities' virtually matched the rate of job growth of producer services, at 13 per cent in total from 1988 to 1995, despite strong fluctuations through the economic and seasonal cycles.

This job growth, so appealing to European governments in the context of large-scale unemployment, results from the favourable interplay of tourist volumes (including the growth of day tourism and second holidays) with particular features of productivity in tourism. Expectation of a 4 per cent per annum growth of the world tourism market from 1995 to 2000 *(Financial Times,* 1 February 1995) is coupled with a relatively poor performance by tourism in productivity which contributes to the increase in jobs. (This combination is the opposite to growth in European manufacturing, which is limited in output but considerable in productivity.) Despite some debate, Medlik (1988) argued that the rate of growth in productivity in tourism was slightly below that of all services or only about half that of the whole economy, and likely therefore to persist at 2 per cent or less per annum. That would be compatible with a continuing annual increase of 2 per cent in employment.

Growth in low income jobs in tourism is underlined by labour shortages and seasonal migration of workers, as from Italy to Switzerland. Tourism generally has difficulty in recruiting and its poor record on training has led to difficulties in retaining staff, with annual turnover rates in the UK varying between 30 and 60 per cent. Even in the last recession, the English Lake District met severe recruitment problems, accentuated by low local unemployment rates, high housing costs and travel-to-work problems. In Great Britain, nominal earnings in hotels and catering for both full-time men and women rose by less than the national average, and tourism is a notable contributor to the UK divergence between rich and poor.

'Multiplier' and 'displacement' effects

Nothing daunted, tourism regularly claims that it generates more jobs than other sectors. For instance, the city of Antwerp occupies a central position in tourist activities in its province of Belgium. In the sixteenth century it was the northern counterpart of Venice and one of the most prosperous cities of Europe. Tourism today plays only a small part in the economy, but its potential led the authorities to commission research. Yzewyn and De Brabander (1992) estimated the province's number of direct jobs to lie between 4,320 to 5,950, while suppliers of goods and services involved a further 1,210 to 1,630, and an additional 2,840 to 3,990 'induced jobs' would result from spending by the workers in the former two categories. Their overall claim was therefore that tourism generated 8,380 to 11,460 jobs, virtually twice the original number, a *'multiplier'* of two.

This may be correct, but is not a procedure normally undertaken by or for other activities. For example, the 'multiplier' from producer services may well be much greater than in tourism due to the higher salaries in the financial and professional sectors. There are just two reasons for including jobs that might result outside the original tourist establishment. The first is that earnings from tourism might be spent much more locally than from other activities, but this is far from being the case. For instance, the colonisation of Italian-speaking Switzerland by hoteliers from the rest of the country resulted in most provisions for hotels in the resorts of Lugano and Locarno being sourced from German-speaking Switzerland, beyond the Alps, and in local culture experiencing only exploitation. The same is true with many hotel TNCs today. It is, however, generally agreed that farm tourism generates more substantial multipliers for the local economy, through the use of local produce, whereas 'timeshare' and second home tourism attracts visitors with fully-laden cars. A study of self-catering and rural attractions in England (PA Cambridge Economic Consultants, 1988) found that while purchase of locally produced goods was low, half the purchases were made through local suppliers, with a much higher proportion at a regional level.

The second aspect is that tourists' own spending may generate new jobs at other existing and new facilities. This may be marked at a local level; the same study, after including this factor, estimated that 9.6 times the original direct jobs in a new facility could be generated in the local area. At the regional scale, however, this was likely to abstract spending from other areas, and the overall ratio was estimated as 6.9. This appears very high compared with work of Johnson and Thomas (1990) in North-East England. The North of England Open Air Museum at Beamish employed 115 direct staff, which attracted fifteen jobs in associated developments, and a total of 178 after allowing for multipliers – but an estimate of jobs lost elsewhere through the *'displacement effect'*, the diversion of tourists from alternative attractions, took them back virtually where they started; at a regional level the estimate was for a net employment gain of 114.

Clearly, any assessment of a prospective tourist establishment must incorporate an estimate for the resulting expansion of jobs at 'tourist-*related* establishments', but equally regional authorities must be aware of new establishments simply taking trade away from others. There is no justification for estimating spending by workers unless comparison is made with other activities in which that is also done. Overall, jobs in tourism tend to be of low quality and are easily exaggerated for new projects. The interest of regional economic development bodies in most of Europe, especially to alleviate unemployment, will in reality lie in actual numbers of direct jobs.

In the light of this, here are some salient examples of development in this growth activity.

The case of France

'Thirty years ago tourism had little impact on the French economy, yet now it has become a major "industry"' (Tuppen, 1991: 191). France has more international tourists than any other country, following a growth rate of 5.7 per cent from 1980 to 1992, as against the European average of 3.5 per cent (World Tourism Organization, 1993). France received 61.4 million tourist visits in 1994 compared with 46.4 million to the USA, its nearest rival, and 43.2 million to Spain. A positive balance of payments in tourism has developed since the late 1970s, and was estimated for 1994 as FFr60 billion, contributing overall 9.1 per cent of total GDP *(Financial Times,* 9 November 1995). Spending by individual foreigners is less, however, than in the US as many come from neighbouring countries for short breaks. Until now, the balance has been helped by the traditional reluctance of French holiday-makers to travel abroad and by their great increase in leisure spending over the last thirty years.

While employment levels in the economy as a whole stagnated, jobs in tourism continued to expand at a substantial rate, estimated at 15 per cent per annum, 1978–88 (Conseil Economique et Social, 1988), and the total number attributable to tourism runs as high as 2 million. However, this must be discounted in most of the ways outlined above. Eighty per cent are in small businesses employing less than ten people and only 40 per cent are permanent. Tuppen (1991) agrees that training and the quality of jobs leave much to be desired, with the benefits being restricted to relatively few areas of the country.

The heaviest concentration of tourists falls on the coastal fringe of southern France, which illustrates the problems of saturation, of building land, access routes and the environment inherent in tourism, and is shared with the adjoining Italian Riviera, the Costa Brava and the Costa del Sol. To relieve some of the pressure, regional development decisions were taken in the 1970s to open up the 'unused' coastline between the Rhône Delta and the Spanish frontier, involving more than FFr6 billion of state investment in the growth of about 30,000 new jobs.

There was also major new investment in Alpine winter resorts. However, some new developments in rural areas proved unsuccessful, and a massive increase of second homes (to at least 2 million) has been contentious. Tuppen (1985) considered the prospects of diversification into urban tourism, being discussed by many towns. This interest was fostered by factors such as 'the generally successful resistance of the tourist industry to the negative impact of the recession . . . , the activity's ability to continue to generate jobs and the perceived importance of its various multiplier effects' (Tuppen, 1985: 11).

Urban tourism accounted for only 8 per cent of tourist activity in France, but Tuppen noted that, in addition to the wealth of attractions in Paris, festivals and cultural facilities attracted considerable numbers of visitors to towns such as Nancy, Besançon, Tours and Bordeaux. Numerous examples of industrial tourism, including nuclear power stations, already existed prior to the establishment of tourist sites at coalmines and at the national car and railway museums at Mulhouse. Despite the financial burdens of exhibition centres and conference halls, these activities are very important as France is ranked second in the world, behind the USA, for the number of international conferences it hosts. Tuppen's overall conclusions are highly relevant to my overall theme:

> With a limited number of exceptions, urban areas do not [in France] represent major centres of tourist activity: nor does tourism appear to represent a leading branch of employment or source of income in many such cases. However, the number of jobs in the tourist industry has displayed a marked tendency to increase, in contrast to other types of activity. Therefore, in large urban areas such as Lyon and Lille where employment is in overall decline, the further development of tourism ought to represent one potentially significant means by which to attempt to stem or reverse this decline.
>
> (Tuppen, 1985: 52)

In the 1990s, more general problems are apparent. It took drastic measures before the large EuroDisney theme park at Marne la Vallée was brought into profit in 1994. Tourism was hit by European recession and the relative strength of the franc, so the number of visitors remained unchanged for three years, 1992–4. In 1995, the onset of Algerian-connected terrorism and reactions to French testing of nuclear weapons in the Pacific reduced visitors, illustrating the vulnerability of tourism to international events. No sector can stand still, and 'members of the industry have long been worried that France's standing as the world's most popular tourist destination may be in jeopardy' (*Financial Times*, 9 November 1995: 11).

The inland movement of British tourism

In contrast with France, the UK has a large deficit in international tourist flows and spending, and has seen a greater shift away from traditional resorts to other areas, stimulated by a worse climate, the growth of second holidays, short breaks and day tourism. In the mid-1980s the government saw the stimulation of tourism as an arm of employment policy, and at least half of all local authorities sought to develop tourism, despairing of any other growth in job numbers.

At the national scale, activity certainly has been growing, only briefly checked by recessions and international events. Roughly speaking, UK international tourism doubled between 1980 and 1995, with overseas visitors increasing from 12.4 to 22.9 million, whereas the exodus of UK residents abroad rose even more, from 17.5 to 40.9 million (*Employment Gazette*, various). However, internal leisure traffic and tourism rose substantially, with day visits exceeding 630 million by 1988/9 (Baty and Richards, 1991). As a result, 'tourism-related' employment (as defined and discussed earlier) increased by a quarter between 1981 and 1989 (Hudson and Townsend, 1992), and by a further 6.5 per cent from 1989 to 1995. Despite falling back in the recession years from 1991 to 1993, tourism and similar services were showing a firmer recovery than was retailing in 1995.

This does not apply to all constituent areas. To British readers it will be clear that their traditional resorts, suffering the impact of foreign holidays, have moved a long way towards other service functions, for instance as retirement, conference, leisure and sometimes office centres, yet failed to throw off their longstanding problems of seasonal unemployment. Urry (1990: 38) argues that 'seaside resorts have become less distinctive because of the widespread deindustrialisation of many towns and cities so that there is no need to escape from them to the contrasting seaside'. With the growth of television, few travel to a resort for live theatre or for the cinema; in Morecambe, for example, Urry reports a major decline in visitors and the closure of the pier and many facilities. In the Thanet resorts of Margate and Ramsgate, many boarding houses have been turned over to retirement and nursing homes and to temporary accommodation for social security claimants.

Such areas have always recorded at least seasonal unemployment (and indeed several were recognised as 'assisted areas' under regional policy 1958–66). In the 1980s they lost hotel employment and saw relative decline in other tourism-related activities. In April 1996, unemployment rates remained at more than 10 per cent of the unemployed in the resort areas of Bideford, Bridlington, Brighton, Bude, Clacton, Dartmouth and Kingsbridge, Folkestone, Hastings, the Isle of Wight, Lancaster and Morecambe, Lowestoft, Minehead, Newquay, Penzance and St Ives, Southend, Thanet, Torbay, Weston-super-Mare and Whitby, most of these 'travel-to-work areas' having qualified as assisted areas in the government's review of regional policy of 1994. Clearly, tourism is not a positive feature in UK seaside resorts.

Where are the gains today? There has been widespread competition by local authorities wishing to promote tourism. A survey by Hudson and Townsend (1992) suggested that only a few areas, mainly in South-East England, saw further scope for tourist promotion. Thus, Oxford, Cambridge and the London Borough of Kensington and Chelsea were awash with tourists and wished to restrict or divert them. While suburban London was developing tourist hotels for the first time, other South-East authorities saw no reason to promote the sector. The most remarkable feature was the number of deindustrialised localities that displayed very thorough and energetic approaches, not only the famous cases such as Liverpool, Wigan and Bradford but also examples in the West Midlands such as The Wrekin, Dudley and Stoke-on Trent or in Scotland (Glasgow and Hamilton).

All have been vying for a total of day spending that had passed the level of £5 billion per year by 1989, in addition to overnight visitors. On the face of it, the big cities won out. With 11 million visitors who spent £140 million in a year over 1988/9, the city of Birmingham was the leading recipient of day-trip income in the country, followed by Manchester (£99 million) and Glasgow (£97million) (including non-routine shopping; Baty and Richards, 1991). These were, however, followed by the small cities of Nottingham and Aberdeen in terms of spending, and by Blackpool, Bristol and Chester in terms of visitor numbers.

Liverpool is a leading example of city promotion. In the face of the worst urban problems in Britain, and with the help of an Urban Development Corporation, it has since the early 1980s developed facilities including a historic dock and a museum dedicated to the Beatles. These attract visitors from as far as Japan and tourism is becoming increasingly evident in Liverpool as an activity that is growing in the face of recession and decline of employment in the car manufacturing and dock industries (Macdonald, 1996). An important finding was that tourism is spread throughout the year, with short-break leisure visitors filling the beds at weekends used by business visitors during the week. This provides one advantage for large cities compared with seaside, rural or historic city tourism. However, the effects on employment are difficult to distinguish in an area of depopulation.

The range of attractions developed all over the country provides severe competition. Many of their visitors (Townsend, 1993) are attracted to rural areas, to historic properties, parks, gardens and wildlife attractions; even factory tourism and steam railways are predominantly located outside urban areas. The principal cities outside London, places such as Liverpool and Glasgow, when taken together do have their full share of visits relative to their total population. It is places like Edinburgh, York or Bath that gain excess tourists compared with their population: cities with an architectural heritage. (They are often relatively strong employment centres as well.)

Job increases from tourism in urban areas have been exaggerated. London, the principal provincial cities and the resorts have all fared worse than the historic cities (Townsend, 1992a). In the terms of Ashworth and Tunbridge (1990), Townsend found that the 'medium-sized multi-functional city', exemplified by regional capitals such as Groningen, Norwich or Portsmouth, has attained more growth than large multi-functional cities or mono-functional tourist-historic cities. A centre such as York remains a long way ahead of a newcomer to tourism such as Bradford.

Above all, it is rural areas that have gained proportionately most jobs (Chapter 10). Outer rural areas have gained from the growth of second homes, second holidays and weekend breaks. The surprise is that the area of most growth, both in the 1980s and in the recovery from 1991 to 1993, should lie in the 'inner rural' areas of the southern half of England. For Bull and Church (1994), the growth of hotel and catering employment here derives from the combination of three economic roles: meeting the needs of tourists, providing a local consumer service (in an area of expanded consumption) and acting as a producer service to business – in an area of expanded business.

POLICY CONCLUSION

Central and local government are prone to exaggerate the scale of jobs and spending in tourism employment and spending, partly because of the genuine research problem of distinguishing tourist activity from local consumption as tourism spreads further from specialised resorts. To Johnson and Thomas (1992) it is not self-evident that governments should be involved (at least to the present extent) in the promotion of tourist establishments – and the UK government followed this logic to the extent of withdrawing Tourism Development Grants in England, but not Scotland or Wales. Research must establish how much promotion of one area or its particular facilities is merely displacing trade from another in a 'zero-sum game'. Which kinds of tourism are sustainable for the longer run, given the industry's tendency to over develop beyond saturation?

In our case study countries, the UK and France, governments have behaved inconsistently over time. Superimposed on national policies are those of the EU (Thomas, 1996), where tourism has been given an increasing role in regional development policy (Akehurst, 1992). In 1984, tourism was specifically identified by the European Regional Development Fund as having much to offer in the creation of jobs, as a labour-intensive activity. The policy framework included enhanced freedom of movement for EU tourists, improved conditions for workers in tourism, the safeguarding of heritage, and latterly the needs for an environmentally responsible tourism policy. There has been no strong policy impact: 'as with national policy, considerable contradictions are potentially inherent within this set of aims' (Shaw and Williams, 1994).

Despite its shortcomings, mass tourism clearly has been the modern salvation of large regions of Spain (Valenzuela, 1991) and other countries. Despite continuing threats to some conventional forms of tourism in the 1990s, it will be the main field for sustaining job levels in a good number of EU regions. In a number of others it may be seen as the salvation of a poor situation, but in the event provide only supplementary jobs: 'there seems little doubt that many local authorities have turned to tourism as a last resort, in the face of falling employment in agriculture and industry and a failure to attract, say, "high-tech" business services' (Hudson and Townsend, 1992: 64). The same authors go on to question the risks of failure, the dangers of different places adopting the same strategy and image, and the implications of external control of staff through national corporations. But . . . there may be no alternative for the individual place.

SUMMARY

- Europe's role as destination for 60 per cent of the world's international tourists is misleading, and is subject to global competition from East Asia.
- Almost 60 per cent of tourism in the EU is within residents' own country, and this is being augmented by the growth of second holidays and weekend breaks.
- Estimates of tourism employment tend to be exaggerated, but in most EU countries it falls within the range of 5 to 7 per cent of total employment, more toward the bottom of the range if other recreation spending is excluded, or if full-time equivalents are used.
- It is the below average growth of productivity that induces job growth, at low wage levels, but it is the fact of growth that attracts politicians' interest.
- There is some justification to the idea of growth in urban tourism; however, it is the historic cities that have more tourists, and rural areas that have enjoyed the greatest proportionate growth.

Part III

CONSEQUENCES ACROSS THE MAP OF THE EUROPEAN UNION

Part II has found many common elements of change in the sectors studied. In particular, elements of flexibilisation and risky or inferior jobs were commonly, but not universally, observed. Restructuring was apparent both within and between sectors as industrial employment shrank and shifted and service jobs grew, sometimes strongly.

Industrial and sectoral restructuring combine in shaping distinct urban and regional outcomes, which will require different policies from the EU, from the super-national to the local level.

What are the policy implications?

The urban/rural dimension is examined in Chapter 10, and regional and industrial policy in Chapter 11.

10

GEOGRAPHICAL TRENDS – TOWARDS RURAL AREAS?

Looking across the map of the EU, both policy and empirical observation can very usefully recognise a distinction between urban and rural areas. This distinction is more marked as regards some fields of change, notably employment change and retirement migration, than for others, and can be extended to provide a gradation of several types from urban to rural. This chapter asks how extensive and systematic the gradation might be. In fact the hypothesis of 'counter-urbanisation' is less regularly applicable on an international basis than in the recent past; notably, half of the 'functional urban regions' of Western Europe have gained population since 1980, while the most remote sub-regions of the EU are still suffering out-migration. This chapter looks at the ingredients of the process of 'urban–rural shift', including the location of small business growth, before concluding with the topic of ongoing EU trends and policy.

INTRODUCTION

The worker of the future seems likely to be a woman, self-employed or part-time and in the service sector – not a man working full-time in a factory. It does not necessarily follow that she is more likely to be making a living in a rural area rather than in a leading city.

That, however, is the case. This is my last major theme: the *tendency* in most countries of Western Europe for a net shift of people and jobs from urban to rural areas in a process described as 'counter-urbanization' (Champion, 1989). After a long period of rural depopulation and poverty, with net migration providing a favourable balance of migrants to the largest cities, the position began to change quite rapidly in the 1970s. Many rural areas went through a reversal of fortunes, so that they started to increase in population size and relative prosperity. Nearer the large urban centres, agriculture was most intensified and the diversification of the rural economy most marked by the time of the 1991 Commission report on *Employment in Europe*:

These areas include rural areas close to the built-up areas and main roads such as South-East England and the Paris–Bonn–Brussels triangle, the lowlands situated close to towns and cities such as East Anglia and the Po Valley, and many regions in the Netherlands, Flanders and Northern Germany. Coastal regions in Southern Spain, the South of France, Italy and Greece, and in the Algarve, Azores, Balearic Islands and Southern England also fall into this category.

(CEC, 1991b: 101)

The quotation provides a useful key to the chapter, but it is important to note that it does not include remoter rural areas of Southern Europe, which still had a large agrarian population and where urbanisation continued later.

The urban–rural shift of population is not always fully based on new jobs in the rural areas. Indeed, the arrival of the new adventitious population is almost a celebration of a 'world after work' which some futurologists proclaim for us all. Certainly, a significant part of the growth of population arises from retirement migration by urban people whose pension and investment income provide for their existence in the countryside. Their choice of area may derive from past holidays in the same locality, and possibly from the ownership of a second home or use of a *gite* there. There are many reports of people advancing the process through 'pre-retirement migration', in which an urban worker gives up the pressures of a city job to take a part-time or self-employed occupation in the rural area, sometimes connected with tourism at the coast or inland. In other cases, the decision to migrate precedes that of setting up a small business. In yet others, the New Age traveller may simply decide that she or he might as well be unemployed in an attractive rural area as elsewhere. Many of these cases reflect a complete re-evaluation of the countryside, with the English in particular cultivating the idea of the 'rural idyll'.

But when leading British politicians (of different parties!) chose to spend their summer holidays in rural properties in the Dordogne (South-West France) or in Tuscany (Central Italy) they are reflecting a common European interest, for the same properties might be rented by, say, Belgian or German visitors, all reinforcing trends toward greener, inland multiple holidays (Chapter 9). Visitors, second homers and in-migrants may not do much for rural shop sales, or offset the concentration of services in the nearest town with a hypermarket (Chapter 8). Nor does the rural idyll appear to have much to do with the growth of producer services in the region of metropolitan areas.

Occupational aspects

Flexible departures from dependence on farming were already long evident. The continental pattern of farming shows many examples of 'pluriactivity'; multiple job-holding has been found to be extremely widespread in a very wide range of rural circumstances (Clout, 1993). Thus the whole notion of the 'farming family' needs to be re-thought, to reveal a very complex array of gendered socio-economic relationships. Part-time farming is a major feature involving, for instance, long-distance commuting in France and Germany to work in manufacturing or service employment. From a study in France, pluriactivity covers a multitude of jobs that relate to rural tourism either on or off the farm (Campagne, 1990), providing not only farm-based accommodation but also farmhouse teas, camping, farmgate sales and sport. The concept of farming as a part-time activity takes a different twist in Greece, Spain and Italy where many farmers with very small holdings work much less than full time on these holdings but have no other source of activity and income. 'Around one-third of farmers in the Community fall into this category, and almost all are found in the Mediterranean countries' (CEC, 1991b: 101). Home-working is expanding rapidly in many such areas.

Newby (1980) argued that farmworkers and their families had been systematically deprived of opportunities, by wealthier groups' manipulation of the British planning system to preserve the countryside (although agriculture, conservation and recreation were themselves at odds with each other). Falling numbers of agricultural employees had kept rural wages the worst in the UK. Now the emphasis in outer rural areas would be placed on the impact of inward migrants and second homes both on the countryside and on the shortage of affordable housing for local people.

There has been a marked upsurge in service sector jobs, both in serving migrants and supporting the dispersal of factories to rural towns and the countryside. For the UK it can be argued that the growth of producer services and tourism in rural sub-regions represents the most important phenomenon in regional development in the whole country. Turning to North American precedent

> For more than two decades the service sector has employed the majority of rural workers. Yet the lion's share of rural research has centered on the agriculture, mining, and manufacturing sectors. Amidst the growing recognition that some service employers are more than passive players in the national economy, a number of researchers have begun to explore what the growth in service jobs means for rural areas.
>
> (Glasmeier and Howland, 1994: 197)

The spread of economic activity back to the countryside was regarded as a great blessing by Fielding (1990: 235): 'Almost every region in Western Europe experienced an increase in female activity rates during the post-war period, but the increases were particularly significant in rural areas.' Men also gained jobs in factories, transport, building and the public sector. Thus, for ordinary working people the range of rural opportunities for making a living greatly benefited from 'counter-urbanisation'. For urban areas the effect was precisely the opposite.

'COUNTER-URBANISATION'

An unexpected employment shift

The recognition of 'counter-urbanisation', first in North America, was somewhat surprising in view of the longstanding, structurally based decline of primary sector employment as a leading cause of rural to urban migration. Counter-urbanisation does not follow directly from the growth of service sector employment. A large part of producer services is oriented to metropolitan areas, as we have seen, although having a propensity to decentralise over wider metropolitan regions. Large retailing complexes are foreign to smaller market towns, and tend to pull spending of all kinds away from village services. Tourism employment has recently swung towards rural areas, but has also grown in cities of all kinds. In a most comprehensive review of the decentralisation of service employment in Europe and North America, it was seen that

> data confirm that the activities which are over-represented in big cities (e.g. business services) tend to show above-average growth rates [in rural areas]. In future, this may be even clearer, since the rather unevenly distributed public services will show decelerating growth. However, the analysis shows that this 'structural' component nowhere has been decisive for the total geographical development in service employment .... Indeed, the majority of countries show a certain decentralisation of service employment in the early 1970s and 1980s. The decisive factor has been a geographical shift out of the big cities, inside many sub-sectors.
>
> (Illeris, 1989: 131)

In West Germany, the same conclusion was prompted by a study of Bade and Kunzmann (1991). Using data for the period 1976–86 they showed not only the familiar finding that southern German metropolitan regions outperformed the northern ones, but that suburbanisation of employment continued and that the periphery showed higher growth rates than the core of cities. Thus, contrary to received opinion, regional development was occurring in ways that were only partly explicable in terms of sectoral change. Likewise, Townsend (1994b) demonstrated how both manufacturing and total employment change in Britain during 1981–9 revolved around systematic departures from forecasts based on the employment structure of different types of sub-region in the early 1980s, with a range of these 'differential shifts' from heavy decline in inner metropolitan areas to high proportionate expansion in outer rural areas.

Counter-urbanisation, how extensive?

'Counter-urbanisation' may have been more extensive in terms of employment than of population, because of the growth of activity rates in rural areas. To population geographers, the phenomenon is defined in terms of a change from population concentration to deconcentration, with the rate of population increase becoming higher in rural than in metropolitan areas, which may indeed suffer depopulation through net outward migration. Though there are problems in agreeing and implementing a precise definition, the observation of a rural 'population turnaround' in the USA of the 1960s triggered off widespread excitement over the possibility of a fundamental change taking place in patterns of population distribution in the age of a motorised society (Champion, 1989). The revival of population growth in the rural USA was confirmed by several studies and shown to represent not merely the extension of existing metropolitan areas but also developments in more rural locations.

As the 1970s progressed, similar trends were identified in a large number of higher income countries, not only in Canada and Australia, but also in smaller and more densely populated countries in western Europe and even in Japan. These changes fitted very neatly with contemporary ideas of a shift from an industrial to a 'post-industrial society'. Vining and Kontuly (1978) first demonstrated the widespread nature of the slowdown of metropolitan growth rates across eighteen countries, including all the northern members of the present EU, together with northern Italy. Although the trend in some EU countries had already passed its peak by the mid-1970s, the 1980s showed it to retain considerable strength. It was very strong in Britain of the 1980s (Champion and Townsend, 1990) and dominated the geography of change through the recession of the early 1990s despite being reduced in volume; an active housing market and climate of business investment appeared to have been prerequisites for a rapid rate of change. The picture generally emerges of a range of different temporal and geographical experiences between countries through the economic cycle, as confirmed by Champion (1994):

> Cochrane and Vining (1988) found fairly consistent signs of renewed metropolitan attractiveness in the late 1970s and early 1980s in peripheral countries of Europe, notably Scandinavia, but much less uniformity in northwest Europe. Whereas a return to faster core growth was observed for the Netherlands and Denmark, the core region in West Germany continued to lose migrants and in France the rate of core population loss was accelerating.
>
> (Champion, 1994: 1504)

Repopulation replaced depopulation across many, but certainly not all, rural areas of western Europe during the 1970s and 1980s (Kayser, 1990).

197

While virtually all 'outer rural' districts of England and Wales have now been caught up in the new trends, this does not apply elsewhere. The somewhat Anglo-American view that stresses past or present counter-urbanisation does not find favour throughout the EU. In Spain, urbanisation remained dominant much longer than elsewhere and depopulation continues in much of the Meseta. Dutch scholars rejected the notion of counter-urbanisation as vague and saw repopulation in parts of The Netherlands as attributable to long-distance commuting by at least one member of the household (Huigen and Volkers, 1991). In many areas, out-migration by agricultural and other manual workers continued as they and their families were forced to seek affordable housing away from gentrified rural areas. Many isolated and very small rural settlements, which lacked reasonable local services and were not perceived as in 'attractive' countryside, continued to experience net out-migration. For example, sections of the Massif Central were less responsive to repopulation than parts of the French Alps.

'Data are sparse and uneven, but it would seem that since the early 1980s, rural repopulation has become less significant in the remoter rural areas of EC12 and has become focused again in countryside within easy access of metropolitan areas' (Clout, 1993: 6/7). Walsh (1991) even writes about *The turnaround of the turnaround in the population of the Republic of Ireland*. Fielding (1990) argues that since the early 1980s the general bias toward growth of rural areas has been replaced by net gains in the rural areas of high growth regions only. Writing on employment, the CEC (1991b) saw the prosperous rural areas near cities (listed above) as only one of three types; the second comprised areas in general decline, with small-holdings on the land and few other opportunities (e.g. the west of Ireland); the third category comprised really marginal areas, often isolated such as the Pyrenees or Greek mountains.

Urban decline?

The general position of urban areas, at least of northern Europe, is that employment trends are more adverse than those suggested by average levels of de-population, with the difference being absorbed by spatially concentrated unemployment and reduced activity rates. Urban depopulation is more gradual than deindustrialisation and related particularly to the densest built-up areas. The scale of economic change in older industrial regions needs little reminder:

> There are sixty regions in the European Community suffering the effects of severe industrial decline, which receive Community assistance under Objective 2 of the Structural Funds .... Over 80% of the population of Objective 2 regions is grouped in urban conurbations, including many of the worst urban blackspots in the Community.
>
> (CEC, 1993c: 3)

In a survey of urban population, Masser *et al.* (1992) note that cities were facing a decline of population for the first time in the 1970s. From Figure 10.1, 'counter or deurbanization tendencies can be found primarily in the most urbanized countries in north west and central Europe, whereas in the Mediterranean (and in eastern Europe) the urbanization process fuelled by urban-to-rural migration prevails' (Masser *et al.* 1992: 111). They see expansion in a limited number of successful international cities, and in many medium-sized towns, but regard 'all other cities' as losers, particularly old industrial cities. They note the deconcentration process within city regions, illustrated by the spread of Paris (Figure 10.2), and relate it to changes in car

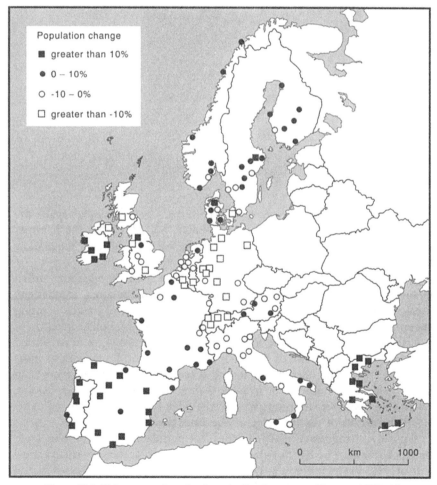

Figure 10.1 Urban growth and urban decline in western Europe, 1970–85
Source: Masser *et al.*, 1992

199

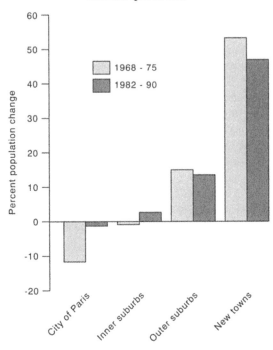

Figure 10.2 Urban–rural shift of population in the Ile de France, 1968–90
Source: Masser *et al.*, 1992

ownership and women's employment patterns. They generally regard the consequences of deconcentration as negative, but admit that the Europe-wide efforts to revitalise inner cities have in some cases been 'remarkably successful' (Masser *et al.* 1992: 114).

Twenty years ago there was an emerging consensus regarding urban change in most of the older advanced economies. Urban cores, or inner city areas, were expected to decline as competition eroded old manufacturing sectors and as suburbs attracted jobs and households. Public investment policies, in employment creation and physical restructuring, were advocated to stabilise the population of the inner areas. This broad framework of ideas has at least been amended. Population loss has slowed, and even reversed, in some major cities. Some inner areas have attracted a middle-class population, while some post-war housing areas on the edge of cities have shared many of the problems of disadvantage of the inner areas.

Figure 10.1 suggested that few cities lost population by more than 1 per cent per annum. The EU population as a whole grew at the very small annual rate of +0.3 per cent between 1978 and 1988. Fielding (1994) stresses that all but ten regions fell between -0.4 per cent and +1.0 per cent. Of the six that fell below -0.4 per cent, two were dominated by declining shipbuilding ports,

Strathclyde and Liguria, while the rest were cities with closely drawn boundaries, such as Hamburg and Bremen (also affected by the extreme trends in shipbuilding). Of the four regions that exceeded 1.0 per cent per annum, only one (Languedoc and Rousillon) was of any significant size. As in the UK, exemplified by the performance of the South West, East Anglia and the East Midlands, therefore, some of the fastest growth was achieved by the smallest regions in terms of population.

Fielding (1994) finds it remarkable that population trends have not been *more* radically affected by changes in the composition of production towards flexibilisation and structural change. For old industrial regions and marginal rural areas, much of the explanation for relative stability lies in the automatic transfer of resources through state social services and welfare systems, and the EU's Common Agricultural Policy. 'This has maintained household income, jobs and services where in other circumstances a downward spiral of economic decline would have set in' (Fielding, 1994: 700). New industrial spaces have been created through the attachment of offices and factories to existing medium-sized towns. Meanwhile, the population decline of the largest cities in the late 1960s and 1970s, even of London, has been checked by the expansion of 'global city' financial services, which has compensated for their deindustrialisation.

The evening out of job opportunities

The urban–rural shift has manifested itself generally in increased activity rates in rural areas, and in greater levels of 'non-employment' in urban areas. As recently as the 1960s, there was still massive rural to urban migration. For instance, in Italy the high rates of rural out-migration of many provinces correlated directly with their dependence on agriculture and its inability to employ a surplus population. 'At mid century over 30 million people were classified as being in agricultural work in the EC12; now the figure is about 9.5 million, representing 7% of the EC workforce and ranging from 2.2% in the UK to 27% in Greece' (Clout, 1993: 7). The number of farms has fallen from 20 million to below 8.8 million. To some extent, the problem of rural overpopulation has been solved, and we can observe return migration of urban surplus labour in the progressive return of former migrants from northern cities to Italy and the rest of Southern Europe whence they originally came (many in retirement, or not otherwise making a salient contribution to local entrepreneurship).

A snapshot of the recent gradient in urban–rural opportunities for making a living in Europe may be taken by reference to new data in CEC, Eurostat (1994), which make a threefold classification of sub-regions between densely, intermediate and thinly populated areas (Figure 10.3). The level of convergence achieved is shown by differences generally of little more than 1 per cent between average rates of employment, economic

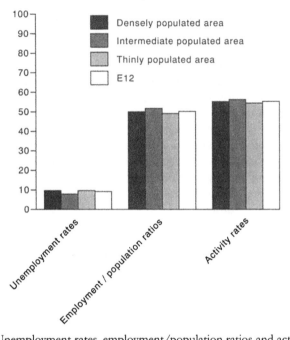

Figure 10.3 Unemployment rates, employment/population ratios and activity rates by
degree of urbanisation, E12, 1992 (per cent)
Source: CEC, Eurostat, 1994

activity and unemployment. In E12 as a whole, the densely populated area
was no longer the most advantaged on any of the three measures. The thinly
populated area suffered the same unemployment rate as the densely, and still
fell 1 percentage point behind as regards levels of employment. The gains
had fallen to the 'intermediate' areas – intermediate both on the diagram and
in geographical position. These had the best scores on all three variables,
fitting with the concept of a spread of prosperity from cities to rural areas
around them, but not to the EU's outer rural areas.

The pattern varied considerably between countries on these measures. The
traditional European pattern was still found with the dense, intermediate and
rural sub-regions ranked in that order. It still prevailed with respect to
employment and activity rates in France and Spain (and to unemployment
rates in Denmark and activity rates in The Netherlands). However, three
countries, Germany, Belgium and Greece, were recording the opposite,
counter-urban pattern in employment and activity, while the UK, France,
Italy, The Netherlands and Greece all recorded such a pattern for unemployment. Patterns were similar for the genders taken separately, although, as
expected from previous chapters, Spain still had only a low proportion, 30.6
per cent, of its rural women reported as in the labour market.

In the case of the UK, it is possible to adopt a much richer elevenfold Census classification of districts (Champion and Townsend, 1990) to assess how employment change and unemployment vary by types of area from London to the outer rural areas. This classification does not supplant by any means the regional differences shown by the waxing and waning of the North–South Divide, but discriminates more powerfully between the experience of areas. Figure 10.4 deals with employment, unemployment (women's and men's) and the convergence of women's activity rates between the full range of urban and rural areas.

In the period from 1981 to 1993 (the longest available for analysis under this system), there was a most remarkable range of experience. Employees in employment fell by more than a quarter (26.8 per cent) in the seven leading provincial cities afflicted by deindustrialisation: Birmingham, Liverpool, Manchester, Sheffield, Leeds, Newcastle-upon-Tyne and Glasgow. At the other extreme, the outer rural areas expanded their employment by nearly a fifth (19.5 per cent): areas such as the South-West Peninsula, rural Wales and the Welsh Marches of England, rural parts of Yorkshire, Lincolnshire and East Anglia, together with the Highlands of Scotland. In between, there was a complete range of experience including significant further decline in London, a moderate reduction of jobs in other cities and industrial areas, and marked expansion in resorts, New Towns and the 'mixed urban–rural areas' that largely represent the Greater South East around London.

As might be expected, this same gradient is apparent in the unemployment rates (Figure 10.4), and in a gradient from more full-time work in urban areas to more part-time in rural. This is somewhat at odds with the continuing characterisation of rural women (farmers' wives and daughters) as suffering marked deprivation with regard to job opportunities (Little, 1991; Townsend, 1991b). The British pattern appears overall to represent a clear example of

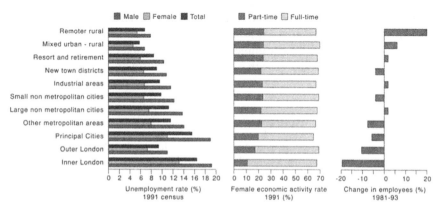

Figure 10.4 The urban–rural range of experience in employment variables, GB
Source: Censuses of Population and Employment

urban–rural shift toward the equalisation of opportunities occurring through the incorporation of women. In net terms, the whole of the expansion of 350,000 jobs (1981–93) in outer rural areas may be attributed to the service sector, but it was remarkable that the reduction of manufacturing jobs was less than 20,000 (4.6 per cent), itself to be ranked as a success story. Due partly to the deindustrialisation of other areas, *rural Britain is now as industrialised, in terms of the share of employment, as the rest of the country.*

THE UNDERLYING PROCESS OF URBAN–RURAL SHIFT

Clearly, job opportunities are only one part of a complex process of change. The overall growth of service employment results in part from the return of population to rural areas. It is one thing to relate together for any one country the statistics for employment, household formation, etc. 'It is, however, quite another thing to be able to identify the underlying factors and even harder to bring forward concrete proof of the precise role which they have played and of their importance relative to each other' (Champion, 1989: 236). No less than seventeen factors are mentioned by Champion. Leading issues comprise the emergence of diseconomies affecting the size of cities, the growth of long-distance commuting, and the contraction in the farming population itself (reducing the stock of potential out-migrants). However, another set of factors relates to the role of government, not only through the conscious promotion of rural development in most countries (Clout, 1993), but also through the past expansion of military sites and the progressive levelling up of rural welfare state spending on education, health, personal services and roads. In the days of deregulation and privatisation in Conservative Britain of the 1980s and 1990s, these policies went into reverse. However, they had greatly facilitated the migration to the countryside of people of all ages, and provided service sector job opportunities in their own right, allowing people to exercise their preference for rural living before and after retirement. Although tourism is important in many specific areas, I believe that the most immediate factors in rural growth are those of retirement migration and growth in both private business and manufacturing employment.

Retirement migration – Paris

The arrival of retired couples in a village will help employment in the general region, although their different demands may destroy most local services – they may use a car to collect all their shopping and other needs in the nearest town, and will fail to support village shops and schools. The growth of retirement migration can be attributed to larger pensions, to the increasing number of years between retirement and death, to expanding car ownership which introduced much of a whole new generation to the countryside, to return migration and in some cases to second homes that later became retirement homes.

The geographical significance of retirement migration is illustrated by the example of the Paris region. Cribier (1982) estimated a rate of retirement migration from Paris of approximately 35,000 per year, with strong effects in stabilising the growth of the city's population and reducing the demand there for social services. Out-migration is most common soon after retirement and affects all social classes, propelled by a lack of things to do and a dislike of unhealthy urban conditions and the prevalence of flats. Dispersal occurred throughout France, with three-quarters of migrants moving more than 120km. This is attributed to the national scale of recruitment to Parisian jobs in the past and to some return migration to areas of upbringing. Upland areas with cold winters tend to be avoided, but the wider Paris Basin and Brittany are favoured, with some spread of the dispersal over time to the Mediterranean and to western coastal and rural areas, including the small towns of the interior. Most retired households are located in a rural commune that is too small to administer rural services to the elderly, and more than one in four is located in a hamlet or isolated settlement. Most households were reported to be happy to have moved, perhaps reflecting the relatively new French critique of the urban tradition.

The growth of rural service businesses?

In addition to serving the augmented population of rural areas, the rapid expansion of service employment includes new business serving wider markets. Some new business is started by working-age, pre-retirement migrants after settling in the rural area of their choice. The time is ripe to extend the scope of services research much further beyond the existing fields of producer services in metropolitan areas:

> The present understanding of the relationship between counterurbanization and economic changes like the shift from manufacturing to services, the growing importance of small and medium-sized firms, or processes of externalisation and internalisation of services is quite limited.
>
> (Jaeger and Duerrenberger, 1991: 117)

These same authors contribute to our understanding of counter-urbanisation in West Germany by showing how consumer, producer and financial services all expanded employment more rapidly, over the years 1976–83, in 'peripheral' rather than 'intermediate' areas, and in 'intermediate' rather than 'metropolitan'.

However, the picture is complex. The evidence from the USA for substantial rural growth in services selling to a non-local market is mixed. Kassab and Porterfield (1991) stress that producer services are growing rapidly in rural areas, about three times as fast as all service employment there but especially near metropolitan areas. However, the number of employees in producer services in rural areas is still quite small.

The location of small business growth in Europe

The relocation of some city-based firms to small towns and rural settlements and the rise of small and medium size industrial enterprises in many stretches of countryside in Western Europe also assisted the trend for repopulation.

(Clout, 1993: 5).

Small business growth and that of rural areas are commonly studied together because of a locational bias in small firm formation. By 1991, some 55 per cent of those employed in the EU's private sector worked in firms with less than 100 employees – 30 per cent in those with less than ten; the percentage differs between 12 per cent in manufacturing and 36 in services (CEC, 1994a). The tendency is to analyse small business growth as a whole, whether in manufacturing or services. Rural manufacturing growth may include many of the most innovative small firms; while a majority of those in services may serve the expanded local population, yet others, most importantly, provide inputs to the national production process, often in knowledge-based services like research, development and planning.

In a major statistical review of small firm births in five EU countries, Reynolds *et al.* (1994) established that manufacturing represented less than a fifth of the births and stock of firms, with local services and building work more than a half. As a result, their modelling showed small firms to be dependent on variables such as population growth:

The highest birth rates in manufacturing in Germany and the United Kingdom occur in rural regions adjacent to the main urban regions. A similar pattern seems to be apparent in the USA .... These rural regions, then, where the economic structure is dominated by a range of smaller firms, have higher rates of manufacturing firm births.

(Reynolds *et al.*, 1994: 451)

Interest in the formation of new, independent small businesses arose from a substantial resurgence in their numbers since the 1960s, after decades of decline. The timing varied, being earliest in the UK and Italy, but it was truly international (Keeble and Wever, 1986). Explanations for the resurgence of new small firms are three in number. Recession-push theorists see the phenomenon as a result of deeper recessions in Europe since the 1960s. They argue that redundancy, fear of unemployment and loss of promotion prospects make more potential entrepreneurs strike out on their own in a recession period than would otherwise be the case. A second view, from Italy (Brusco, 1986), suggests that substantial income growth can lead to a sophisticated widening of consumer tastes and the growth of flexible small firms to accommodate the demand. Lastly, the 'technological change' towards a micro-electronics-based information society is also seen to encourage a more decentralised, flexible pattern of production.

Such factors have encouraged average rates of new firm formation that are roughly similar across six present members of the EU. For a range of dates between 1981 and 1991, annual firm births per 10,000 persons range from 33 to 144 for all sectors, and 7 to 28 for manufacturing (Reynolds *et al.*, 1994). Ratios showed similar variations across all countries, with the most 'fertile' regions having annual new firm birthrates that were two to four times higher than the least fertile regions. 'The underlying processes affecting new firm births at the regional level appear uniform across countries' (Reynolds *et al.*, 1994: 443), and are related to growth of population and demand, accompanied by an existing density of firms. Very significantly, neither high unemployment nor supportive government policies made much impact. The prospects for areas of high unemployment, many of them urban, from this source of growth are therefore few.

UK cases

The UK provides a full spectrum of area types from the most urban to the most rural (Figure 10.4), and results of extensive interviews with small firms (for example Curran and Blackburn, 1994; Keeble and Tyler, 1995).

In a review of data across the spectrum of area types, Townsend (1994b) posited an urban–rural cycle extending through the recession and expansion period of the 1980s. The early part of the 1980s showed recovery, extending well beyond what could be expected on structural grounds, beginning in the least urbanised areas of the South at a time of continuing deterioration in London and other cities of the South. Northern cities showed a neutral performance that contrasted with reindustrialisation in other areas and proportionately strong growth of manufacturing in rural areas. Even with a further recession in the early 1990s, manufacturing jobs in 'remoter, mainly rural areas' had still declined only marginally by 1991.

The net growth of small business provides the immediate explanation for this most unusual achievement. The deeper explanation lies not in government policy for Rural Development Areas (Rural Development Commission, 1993), but in a set of advantages identified in surveys of firms. Curran and Blackburn (1994) studied over 400 businesses in seven contrasting localities in the UK. Their findings challenge accepted views that close links exist between small businesses and their local economies and that local economic networks are important for small businesses. Curran and Blackburn argue that small firm–large firm links are less close (and less and less local) than is often assumed: this brings into question the applicability of the Italian model of 'industrial districts' that are idealised in some interpretations of flexibilisation .

Such long-distance relationships enable firms to profit from the net advantages of a rural location. Keeble and Tyler (1995) surveyed over 1,100 businesses in urban and rural areas of the UK. They found the leading success factor in rural areas, mentioned by 80 per cent of respondents, to be the attractive living environment, followed by good labour relations and lower wage costs. Significant minorities mentioned disadvantages: poorer rural proximity to suppliers and customers, and thinner rural availability of skilled labour and business advice.

Does the evening out of job opportunities between rural and urban areas represent exploitation of the rural worker? Moving from old factories with the most highly unionised workforces, Massey and Meegan (1988) suggest that companies prefer to locate in rural areas or small towns where there was thought to be a good supply of 'green', un-unionised and relatively cheaper labour with little tradition of work in manufacturing. Basic manual wages certainly provide the lowest incomes in rural counties (Champion and Townsend, 1990), an influence persisting from the agricultural past. Church and Stevens (1994) suggest from new data that variations in union density between urban and rural areas are only moderate, and that companies investing in small towns and rural locations would not necessarily escape trade unions.

Counter-urbanisation does, however, represent the march of technical change and flexibilisation. A Cambridge team explored spatial variations in technological and other innovative activity between urban and rural SMEs (Small and Medium Sized Enterprises). The highest rates of innovation, R&D and technological expertise were all achieved by rural enterprises. 'It would appear likely that this represents one major reason for the better employment and export performance of rural SMEs noted earlier, and indeed for the general urban–rural employment shift in Britain since the 1970s' (Small Business Research Centre, University of Cambridge, 1992).

Keeble and Tyler (1995) go further in relating innovation to its environment, as both attracting innovative migrants in the first place and supporting their enterprises once established. These dynamic effects are fuelled by rising real household income and a proliferation of market niches for specialised and customised products and services as posited in flexibilisation theory (Hirst and Zeitlin, 1992). The contrast with urban areas is extreme, and suggests 'that economic revival of the production base of less favoured areas may only occur in the longer term if these areas can provide relatively attractive *environments* to those who wish to invest in order to undertake the production of goods and services' (Keeble and Tyler, 1995: 995, emphasis added).

PROSPECTS FOR RURAL AREAS

European policy for rural areas

Policy for rural areas was a complex matter even before some of them, mainly in the UK or northern Europe, achieved a 'turnaround'. 'Each member state of the EC has devised and implemented its own, often very complicated, version of rural policy' (Clout, 1993: 9), resulting from specific circumstances of rural poverty, the political muscle of farming lobbies, the need for regional and sub-regional physical planning and the activity of regional bodies in federal states. The Common Agricultural Policy itself had a great impact on rural areas. Beyond that, three phases of activity may be identified across western Europe. The first aimed to enhance living standards of rural families through land reform and modernisation, notably in the Mediterranean. The second sought to provide non-agricultural work, first in manufacturing and later in recreation and tourism. More recently the emphasis has been on the local community's participation in locally generated grassroots schemes.

EU policy has reflected all these strands, achieved through financing development programmes negotiated jointly with member states and regions (CEC, 1995c). Rural areas are included in the EU's first priority areas for regional assistance ('Objective 1', see Chapter 11), but are specifically the target of 'Objective 5b', which focuses on the economic diversification of rural areas, including 8.2 per cent of the EU population. The Commission's guidelines on employment promotion, interestingly, are set out in the following order: support for SMEs and the craft industry sector (backed up by studies such as CEC, 1995d); tourism; environmental protection (which may yield productive investment, e.g. in water control); investment in infrastructure; and improved co-ordination. More innovative is the approach of LEADER, dating from 1991 and concerned with support for local action groups and with public-private partnerships active in, for instance, training or community services.

The EU itself stresses the limited resources available for 5b areas, and it would seem that action there achieves significance only in the context of low population numbers (as in the growth of new firms in mid-Wales). My District Council experience in the Pennines at the date of writing is that these funds are hard to access. The Highlands and Islands of Scotland (Black and Conway, 1995) show some benefits to be gained from the LEADER bottom-up approach to rural development, but the EU scheme did not provide funding for long enough to animate local inputs or to network ideas in Europe.

Prospects for urban–rural change

Even if it does not benefit the most remote rural and agricultural areas, will the balance of advantage continue to work against urban areas, perhaps in favour of Eurostat's 'intermediate' areas (Figure 10.3)? Masser *et al.* (1992: 118) posit a 'growth scenario' (based on observations in cities like London), under which industrial parks near motorway exits contribute to sprawl which devastates the countryside around conurbations and resorts but benefits well-connected, freestanding towns – though one could 'surmise that more peripheral rural regions will suffer from depopulation and neglect'. Their 'equity scenario' envisages a balance of urban, suburban and rural life opportunities constructed by successful policy. Their 'environment scenario' imagines how city and countryside might look in 2020 if resource conservation were the main goal of settlement planning, resulting in widely distributed, low density developments, with dispersed services and employment.

The experts whom they assembled saw the growth scenario as the most likely future, with variations between different parts of Europe. For them, it was possible to speak both of a growing polarisation at the highest level in Europe, and of a diminishing distinction between city and countryside toward a 'rurban' continuum around city regions.

In the USA, selective urban recovery is appearing. Noyelle and Stanback (1984) had already predicted that the enhanced role of the service sector would reverse the redistribution of population, especially with producer services benefiting so much from agglomeration. Frey (1993) was convinced that a new urban revival in the USA represented a return to urbanisation after the counter-urbanisation of the 1970s, although these were changes that would remain fluid just because of the continuing pace of employment change.

In western Europe as a whole, Fielding (1994) found that structural change had been accomplished surprisingly simply, by the attachment of new offices and factories to existing towns of different sizes. Sectoral restructuring and flexibilisation had been accomplished without major change to the settlement system. The population decline of the largest metropolitan areas had been arrested, or reversed, as the accretion of 'global city' functions to these areas had already compensated for their deindustrialisation.

The urban–rural shift is less deep-rooted and persistent elsewhere than it appears to be in Britain. Nor indeed does it enjoy the same regularity as other themes of this book: the swing towards service, women's and part-time employment. Indeed, under the environment scenario, the tensions evident in rural areas of Britain and elsewhere already show that repopulation of some rural areas has passed a critical point. Communities, agricultural land and conservation areas are under threat from house-building, tourism, recreation and new kinds of sport. Young and other low income people are priced out of the housing market by second home purchase and newcomers, while planning restrictions place a limit on the expansion of new businesses. Taking this view, the Rural Development Commission pressed for a more favourable view of business growth in the government's 'Rural White Paper' (Department of the Environment, 1995). The British government failed to resolve conflicts over policy for rural England, or indeed to provide any fresh resources. Certain guidelines were, however, revised in favour of the reuse of rural buildings for business rather than residential use, and a new 'rural business use' was identified. All appear minor influences compared with the preferences of entrepreneurs within the overall system of a nation's spatial economy.

We might conclude then that the urban–rural drift will continue to wax and wane across Europe according to the pulse of the economy and to secular changes in the role of individual sectors, notably the role of producer services in restoring some centralisation.

SUMMARY

- Employment of many kinds has been involved in the complex relationships prompting a reversal of the directions of migration in favour of rural areas, though excluding the more remote areas of Southern Europe.
- The urban–rural shift has remained less constant in many countries than in the UK; in particular, half of the functional urban regions of Europe have gained population since 1980.
- Although retirement of people from areas of higher to lower density of population is a major factor in the growth of rural areas, so too is the bias in the formation of small businesses against more urban areas, based on the growth of both population and demand.
- EU policy for rural areas, other than the Common Agricultural Policy, is comparatively poorly resourced, but, as in the UK, is breaking new ground in terms of community involvement.

11

CONCLUSIONS, AND
IMPLICATIONS FOR
EU POLICY

All our futures in the EU are intrinsically bound up with continuing problems of unemployment and of the quality of jobs. We do not yet have an alternative to high levels of employment in paid work to meet our psychological needs and to distribute wealth in our society. We can produce considerable wealth, but without jobs this will leave many in poverty, deprivation and social exclusion. What then can be done, through European, regional and place-specific policies, to alleviate these problems in the short and medium term? The long term will depend on a new kind of society.

This chapter provides a summary of all the main findings of the book about employment change, with particular reference to our two main themes, sectoral change and policies, and the field of 'flexibilisation'. It finds cause for fresh alarm not only from the persistent nature of unemployment over much of the EU, but also from the very effects of the EU's own drive for monetary integration. A concern for traditional spatial policies is being overtaken by problems of cohesion and risk of deeper social unrest in much of the EU area. A full review of the now conventional list of answers to the employment problem leads us to place heavy weight on the unrealised possibilities of agreement for reducing working hours, but also to review more radical possibilities. After a decade in which I have seen a Conservative government encourage rather than stop small scale intervention in support of village community projects, I will support Brussels' concepts for local initiative in the voluntary and service sectors.

A RESUME

In Part II, I discussed change in leading sectors of work individually, in order to establish their roles both in producing unemployment and low pay and offsetting these problems.

Sectoral change has accelerated since 1971. The loss of jobs in the primary and manufacturing fields has been more severe in the EU than in North America, and the growth of the service sector, although common to all areas of the EU, has failed to offset it. Very different countries in Europe resemble each other on these dimensions and none has a good prospect for jobs. Even the positive effect of the Scandinavian welfare state on welfare jobs (Esping-Anderson 1990) has become outdated: in the years preceding the entry of Sweden and Finland to the EU, the viability of Scandinavia's high income economies was in doubt and there were significant losses of service sector jobs (p. 151). Although in most areas of Europe tourism and leisure jobs are expected to diversify and grow (save in traditional holiday resorts), banking jobs have passed their peak in richer countries (p. 151), and so, in the UK, The Netherlands and Denmark had women's and part-timers' jobs in retailing (p. 163). (Growth in women's jobs in the USA is also expected to moderate.) Net job growth in manufacturing is limited to the environs of particular new plants, as in southern Spain, although it is rural areas of northern Europe that come closest to sustaining their totals of factory jobs (p. 210). Sectoral changes differ between areas, but growth in productivity has left many places dependent on a more precarious mix of jobs.

Some commonly described threats to the established order nevertheless contain an element of journalistic exaggeration. Jobs are not draining out of Europe. Comparatively few reports exist of the direct transfer of whole factories from high to low income countries, although I believe they have increased in number. Even the concept of 'a new globalisation' appears exaggerated when present links are compared with those of Europe before the First World War. Many of today's interconnections take place in and have grown through trade with the rest of the industrialised world. The threat of, say, substantial software imports from the Indian sub-continent is limited to particular sectors and sizes of contract. The US concept of post-industrial society is not justified on economic grounds, although it is the experience of workers.

'Flexibilisation' and the concomitant increase in insecure, low paid work are firmly established as a long-term trend. Its impact, notably in retailing, can be great, but its pace and nature have been exaggerated as regards most of its elements and relationships. Most countries can be seen to be following the USA in the general direction of flexibilisation, but the starting point varies between countries and regions by virtue of their different cultural histories and legal traditions. The UK and Denmark are extreme in their degree of reliance on *part-time* workers, and in the E15 such workers form a small and inconsistent part of any general trend. Numbers of *temporary* workers remain small in most countries and are increasing only slowly. The literature on the *informal* sector, home-working and tele-working shows them to be still minority activities (p. 127). Tourism does demonstrate a funda- mental reliance on *insecure* workers, but this is increasingly alleviated by the expansion of winter breaks and inland locations (p. 188). Producer services do employ considerable numbers of relatively *low paid* women, but mainly on permanent, full-time contracts. Many writers ascribe the growth of producer services to the *contracting out* of work by industrial firms; in reality, work comes from a much wider range of sources (p. 136).

The concentration of high unemployment in particular parts of the EU is explained in part by the distribution of these sectoral and occupational trends. Although a 'polarisation' of jobs is clearly evident within, say, retailing, talk of 'polarisation' of incomes within metropolitan areas as a whole is more marked in the UK than in most other member states. In today's conditions of high unemployment, the biggest contrast is between 'job-rich' and 'job-poor' households. Metropolitan and urban areas generally have greater concentrations of unemployment than surrounding 'intermediate' areas, despite their retention of most of the core offices of producer services and the corporate sector. 'Thinly-populated areas' of the EU have the same average rates of unemployment as urban areas. Generalisations about 'counter-urbanisation' may have less validity over time and space than has been thought.

Regional differences in employment levels have increased over time, for specific reasons in a unified Germany but for more general reasons in Italy and in both rural and urban Southern Europe at large, which share both higher rates of population growth and the difficulties of attracting investment into the periphery of the EU. At the same time, employment structures are distinctive and flexibility arises from a balance of self-employed and family patterns of enterprise, whether in farming, retailing or tourism, which is very different from the north and west of Europe.

NEW ALARM

What will be the overall balance of job seekers and opportunities in the EU? What will be the imbalance in individual places? Could such imbalances be seen as representing, in EU language, a threat to the 'cohesion' of the Community? The process of unification and harmonisation is itself seen as a possible cause of disruption and regional unemployment.

Such problems are neither new nor specific to the EU. Great disruption is being risked in the establishment of continental trading blocs elsewhere, for instance in the North American Free Trade Area (NAFTA, Chapter 2). One of Europe's greatest regional problems, that of Southern Italy (Chapter 1), dates back to the withdrawal of tariff protection upon the unification of Italy. Until 1860, the Kingdom of Naples controlled the whole of the South, and as an independent European country had its own frontier posts and a wealthy court, with a full range of craft industries. Upon its amalgamation into a unified Italy, the South lost the protection of these tariffs and found itself subject to the policies and interests of the northern leaders of unification. This not only put many southern craft businesses out of work, as northern goods became more cheaply available in the South, but began an outflow of investment and skills to the North which contributed cumulatively to the post-war situation of, for example, the Region of Campania (p. 19).

The CEC has long identified a political need to contain such imbalances through regional policies, but their effect was not sustained through the 1980s (Chapter 3). 'Regional inequalities have started to increase again and by 1990 they were as high as in 1970' (CEC, 1994b: 238). Many authors witnessed increased social as well as spatial differentiation across the EU even before the later stages of integration: Kesteloot (1995), for example, identifies the immigrant ghettoes of Brussels as housing the very people who might be employed in doing the menial photocopying work at the European Commission headquarters. Other contributors to Hadjimichalis and Sadler (1995) stress the vulnerable position of different social groups, such as guestworkers, contract workers and, in Southern Europe, women, who all suffer from broader changes in welfare systems. Many such concerns were submerged under the general 'Europhoria' of the late 1980s and the early 1990s.

CEC policies themselves as a threat to employment

In the 1990s the CEC has itself identified the threat of further integration to the 'cohesion' of the Union and member societies. European economic integration seems to pose particularly severe problems of industrial adaptation for weaker regions. Enhanced structural policies are regarded as necessary to allow weaker regions to adapt to new competitive conditions. Writers such as Amin and Tomaney (1995) argue that the resulting measures are insufficient to counter the powerful centralising tendencies to be released by monetary union. As viewed by the British Labour Party, at least in the past, the EU might be seen as a club for TNCs, with new social and spatial policies as the fig leaf to suggest political protection.

By 1996 the persistence of high levels of unemployment (the average across the whole EU being 11 per cent throughout 1995) contributed to increased concern over EU and national policies. A particular underlying factor was for member states to start preparing for monetary union, by trying to satisfy the strict 'convergence criteria'; the deflationary effects on employment were a principal factor in retaining minority opposition in the British Labour Party to further measures of unification. Several countries, interested in joining a single European currency, decided to employ restrictive financial policies, which caused extensive public sector strikes in France in 1995 and demonstrations in Germany, unprecedented for forty years, in 1996.

Belgium did not show such signs of unrest but shared a dismal economic performance during 1995, with Belgian unemployment throughout close to 10 per cent of the workforce on international definitions:

For such a European nation – multinational, geographically central and

home to the EU's institutions – the figures highlight a sad irony: name-
ly that the country's integrationist ambitions are taking a heavy toll on
the economy . . . . The prime minister . . . has vowed that Belgium will
be among the first countries to replace its national currency with the
Euro.

But in order for the dream to be fulfilled the country has had to stick
to a series of tough austerity measures. For example, it has frozen
wages since 1993 and cut spending on health and social services.

(Tucker and Tett, 1996: 3)

The new cause of reduced government spending is that a country is allowed
to carry only a minimum of annual debt into the proposed common currency. Belgium will contain its budget deficit to the levels required for entry
into the common currency, but was also affected by adverse trends in the
European economies generally. The response of French and other finance
ministers was to insist that monetary union would go ahead on schedule in
1999. But for those who had hoped to start full union then, the downturn of
1996 could not have come at a worse time.

The poor economic performance of the 1990s comes on top of a longstanding failure to generate new jobs. Six million EU jobs were lost between
1991 and 1994, twice the size of any contraction of the workforce in
employment since before the Second World War. More than 1.7 million jobs
were lost in Italy, 900,000 in the UK and 800,000 in Spain during that period.
There was a dramatic fall of nearly 3 million in the number of young people
under 25 in the workforce. Between 1991 and 1995, GDP of the EU
expanded by only 1.6 per cent per year, insufficient to match the regular
growth of productivity, and therefore the EU standardised rate of
unemployment rose, from 8.8 to 11.0 per cent, with particularly strong
increases in Germany, Sweden and Finland. Sweden saw for the first time a
general decline in jobs in some parts of services (Sjoholt, 1994) and warned
that unless the fight against unemployment were given fresh impetus, then
Europe's social and economic stability would be at risk.

In early 1996, German unemployment passed 4 million for the first time
since the Second World War, and the rapidly deteriorating labour market
position became the leading political issue, emphasised by heavy redundancies, such as 9,000 in Daimler-Benz Aerospace. This was the more remarkable because production continued to recover, illustrating Dreze and Bean's
point (1990), that the process of wage formation process in Europe makes
growth in jobs dependent on sustained growth in output. Most employers
were concerned to cut costs but among the public,

most Germans are opposed to the US-style labour market deregulation to create low-paid jobs – this is viewed merely as a form of
hidden unemployment. The phenomenon of the working poor has
received much publicity in Germany, where television programmes

216

show shoe cleaners or bag-fillers at US supermarket check-outs working for low wages.

There is also a rejection of the idea that jobs lost in manufacturing can be made up in service industries .... But it is unlikely that the country will be able to create the jobs in manufacturing to take up the growing numbers of unemployed without more painful adjustments than are now on the political agenda.

(Muenchau, 1996: 15)

Any likely upturn would be unlikely to solve the German unemployment problem. According to Norman (1995), only a third to a half of German unemployment was of a nature that could be helped by economic growth. German industry had used the previous recession to shed labour, with some continuing relocation to other countries by international business. He concluded that Germany's job crisis was increasingly structural, rooted in industrial change and training.

That too was the view of the CEC when the issues came to be debated, for example at an international 'summit' meeting at Lille in 1996. A co-ordinated stimulus to economic growth would reduce EU unemployment by only 1.6 percentage points; part of the structurally unemployed group could be reabsorbed through training measures, which have been too weak to keep up with technological demands as the economy moves rapidly to knowledge-based production. However, this would still leave a substantial number of people facing the chance of 'social exclusion' (*Financial Times,* 1 April 1996: 4) and concentrate the risk of unemployment in cities (Symes, 1995).

SECTORAL CHANGE AND POLICIES

This section introduces a final overview of sectoral change, with an introduction to EU sectoral policies. By setting out data up to 1994, Table 11.1 demonstrates the continued role of many points in Part II for employment levels, notably the transfer of jobs to services, still more from agriculture than from industry:

The most striking difference between the Community and the other economies – specifically the US and Japan – is not in terms of the sectors in which employment has risen or fallen, but in the scale of the job losses which have occurred in the declining sectors.

(CEC, 1994a: 171)

All net additions to jobs in the EU were in services, often as fast as in Japan. The difference lay in the other sectors which in Japan lost much fewer jobs.

Table 11.1 Employment change by economic sector, 1987–94; eleven countries (E12 less Germany)

	Total		Change, 1987–94	
	1987	*1994*	*nos.*	*%*
Agriculture	8,725	6,339	−2,386	−23.9
Industry	23,356	21,296	−2,060	−8.8
Construction	7,762	7,580	−182	−1.9
Services, incl.	59,112	66,184	+7,072	+9.6
distributive trades				
and hotels	19,318	20,166	+848	+3.6
Transport and				
communication	5,825	6,157	+332	+5.7
Banking, finance				
and insurance	6,954	10,646	+3,692	+41.2
Public administration	7,438	7,717	+279	+2.8
Other services	19,578	21,499	+1,921	+8.1
Total	99,351	101,718	+2,367	+1.9

Note: Germany is excluded because, for these data, the source provides no consistent series before and after reunification.
Source: CEC, Eurostat, 1993c, 1996c

Agriculture

European agriculture offers an illustration of the 'dysfunctionality' of increased labour productivity. The social and political costs of the increased unemployment which is likely to be an effect of relaxed protection may serve, in effect, as an argument for the much criticized Common Agricultural Policy (CAP).

(Ahnstrom, 1990: 179)

It is necessary to relate the CAP to our theme of employment. The aims of the CAP, as part of the original agreement for the Common Market of six nations, were to guarantee a fair standard of living through stabilising prices and ensuring productive use of labour. The CAP is generally seen as being too successful, particularly in generating butter mountains, but it also offset profound effects on jobs from continuing mechanisation and changes in farm size. Table 11.1 reflects substantial continuing declines in employment in agriculture in all countries except the UK, Belgium and The Netherlands. This decline is expected to continue, with agriculture falling to 4.3 per cent of EU employment by the year 2010 (Cole and Cole, 1993). The UK and The Netherlands are among those countries that have argued most strongly for reform of the CAP, and the CEC has considered price cuts of up to 35 per cent in some sectors. To summarise, policy changes will play their part in a large and continuing loss of jobs from the land which will be proportionately greatest in Greece and Spain, and which will therefore contribute to

218

inequality in some of the most peripheral and thinly populated parts of the EU.

Manufacturing

There is little reason to expect a future common manufacturing policy to protect manufacturing in the way farming is protected by the CAP. But the creation of a 'true European market' may provide for a substitute that shields particular manufacturing industries from the competition of foreign producers.

(Ahnstrom, 1990: 180)

The total EU finance available to industry amounts to no more than 10 per cent of the budget, whereas agriculture receives more than 50 per cent. Given the importance of industry, and the significance of industrial job losses (Table 11.1), it is surprising that the EU does not have a single, explicit overall policy for industry. The Single Market programme itself, with its emphasis on EU industry and goods, is one aspect of policy aimed at promoting the fortunes of business with a view to improving its competitiveness. Specific measures are focused on key industrial sectors, as in longstanding assistance in coal and steel, in restructuring the iron and steel and shipbuilding industries, in particular trade agreements, as in the import of textiles or motor vehicles from the Far East, and in the funding of primary and applied research in new technology.

'Factory closures and declining manufacturing employment are no longer restricted to a small number of traditional industrial "basins", but are now features of numerous localities' (Tuppen and Thompson, 1994: 159). The generality of deindustrialisation (except in a few semi-rural localities of Southern Europe that attracted large-scale plants) was especially marked in France and the UK in the period of Table 11.1 (Chapter 4). The recession of 1990–2 reversed such growth in industrial jobs as had occurred, and growth was thereafter restricted to special cases such as instrument engineering and biotechnology.

Services

By contrast, service trades were generally the subject only of much more recent and/or weak policies, as in the programme for legal harmonisation of international trade in producer services or the adoption of common standards in tourism as between member states. For all that, Table 11.1 clearly shows that massive growth in 'banking, finance and insurance' (producer services, Chapter 7) far outstripped the effects of growth in tourist jobs (Chapter 9) and the small job growth resulting from automation of retailing jobs (Chapter 8) in 'distribution, trade and hotels'. Significant job growth in services occurred in the first half of the 1990s despite the effects of recession.

Most expanding branches of services involve either the provision of collective services, such as water, health care and sanitation, education and public administration, and/or those with a significant amount of central, regional and local government involvement, such as recreational services or post and telecommunications. The specific mixture of service jobs varies over time and to some extent between countries; for example, mergers and branch closures in banking and insurance themselves contributed to reaching a plateau in the overall level of jobs after the net losses of the recession. Illeris (1991a: 1) argued that there were 'many roads to a service society', with the welfare state prominent in Sweden, and drew a strong contrast between a long upturn in service jobs in France and a level of comparatively recent under-provision in Western Germany, which is confirmed as real by CEC (1994a). However, the structure of growth is remarkably consistent: 'the general areas of job growth have been much the same in the Community as in the US and Japan' (CEC, 1994a: 22).

FLEXIBILISATION – LESS CONSISTENT

Taking a final view of flexibilisation, there is now great policy interest from some quarters in the EU in less rigid protection of the full-time employee. The expansion of other forms of employment makes inroads into the dominance of conventional forms of work and, of itself, raises other policy questions, particularly the risk of a possible *divergence* or *polarisation* of incomes and job opportunities. As was clear in the chapters on producer services, tourism and retailing, while the growth of services is leading to an expansion in the demand for highly-skilled labour, it is also creating a significant number of low skilled jobs:

> Overall, however, there seems to be a shift in occupational structure towards more skilled jobs. Between 1983 and 1991, the numbers classified as professional and technical workers in the Community expanded by over 2½% a year . . . .
>
> Professional and technical jobs, which accounted for only 15% of the total in 1983, were responsible for 40% of the overall rise in employment.
>
> (CEC, 1994a: 22)

Compared with the clear directions of structural and occupational change and of feminisation, the prominent issue of 'flexibilisation' proves increasingly complex. The much vaunted spread of non-conventional jobs appears much less of a novelty and considerably more patchy and slow moving than in media accounts, and indeed in the research questions with which this book began. Although difficult to measure, the relevant trends are depicted in Table 11.2, which considers the expanded E15 up to 1994.

There is greater variety in trends and in proportions here than in other tables. Taking the sum of self-employment, temporary workers and part-time workers as a surrogate for 'non-conventional employment' (inevitably involving some double-counting), they already represented 37.4 per cent of E15 employment by the year 1985 and then increased at a modest rate to achieve 41.9 per cent by 1995. At the same time the proportion of women among the employed of E15 increased from 38.2 to 41.3 per cent. Although the self-employed were predominantly men, the activity rate for men of working age in the EU fell considerably from 1990 to 1994. The statistical shift toward part-time work may be exaggerated. The proportion of people normally working 10 to 29 hours per week in the Community increased by much less than the proportion classified as in part-time employment between 1983 and 1992. And while the number of workers with temporary contracts was growing, they still constituted only one in nine of the work-force in 1994.

Variation between countries in Table 11.2 applies to directions of change as well as to absolute levels. While Spain records a full third of employees on such contracts by 1994, this was totally exceptional, and in seven countries

Table 11.2 The modest growth of recorded forms of flexible employment, 1985–94, E15 (percentages of total employment)

	Self-employed		Fixed-term contracts		Part-time employment	
	1985	1994	1985	1994	1985	1994
E15	15.4	15.6	9.1	11.0	12.5	15.3
Denmark	9.9	8.4	12.3	11.9	24.3	21.2
Finland	14.8	15.5	10.5	12.9	8.2	8.4
Sweden	9.0	10.3[1]	11.9	11.5[1]	25.3	24.9[1]
Austria	15.3	14.5[1]	[2]	[2]	7.0	8.7[1]
Belgium	15.9	15.3	6.9	5.1	8.6	12.8
France	12.6	11.8	4.7	10.9	10.9	14.9
Germany	9.2	9.3	9.8	10.2	12.8	15.8
Ireland	21.5	21.6	7.3	9.4	6.4	10.8
Luxembourg	9.4	9.7	4.7	2.7	7.2	7.9
Netherlands	9.1	11.1	7.5	10.9	22.4	36.4
United Kingdom	11.4	12.9	7.0	6.3	20.9	23.8
Greece	36.0	34.4	21.2	10.3	5.3	4.8
Italy	24.1	24.1	3.7	6.1	5.3	6.2
Portugal	26.2[3]	25.2	14.7[3]	9.3	6.0[3]	8.0
Spain	22.4[3]	22.1	15.6[4]	33.6	5.8[4]	6.9

Note: 1 1993; 2 not available; 3 1986; 4 1987
Source: CEC, 1995b

temporary employment actually declined. Self-employment was highest in Southern Europe, in farming, retailing and tourism, and was tending to decline from these high levels, while the UK was untypical in its scale of movement toward greater self-employment (see above). High levels of part-time working in The Netherlands are confirmed by the CEC (1994a, p.117), and linked to the very low levels of women's participation in the 1960s, since when part-time work has become very much an accepted feature, for some men as well as many women, and has state support. Moving further north, however, we find evidence from Denmark and Sweden of part-time employment passing its peak (Table 11.2).

A UK study, using slightly later data, argues that a realistic picture of work remains elusive, as the labour market is growing increasingly complex and difficult to interpret. Business Strategies (1996) found that while some change was occurring, it was gradual and not necessarily permanent. The proportion of employees in self-employed, temporary and part-time work had not changed significantly in the previous ten years and was not expected to change greatly in the ensuing ten. Beatson (1995), however, concluded that the UK labour market had become more flexible, more evidently as regards status and hours than as regards wages:

> In a number of EU member states, flexibility tends to be limited at the micro level. Labour market regulations may be an important factor. The evidence suggests that this may not always be compensated for by flexible wages at the micro level.
>
> The UK appears to be in an intermediate position, with some features of its labour market resembling those of the USA, others those of its EU partners. However, the UK has probably moved closer to a US-style labour market since the end of the 1970s.
>
> (Beatson, 1995: 64)

The UK does share with the USA a growing fear about security at work. This is apparently out of all proportion to recorded change, especially among white collar workers. There has been a small reduction in men's average duration of jobs, especially outside the age range from 25 to 54. However, 'the message on job insecurity in the UK and the US is the same. Whatever the woes of core white-collar workers, they are dwarfed by the deterioration of the labour market outside these professions' (Flanders, 1996: 19). Up to 1996 the UK has seen the failure of employment to recover to anything like its pre-recession level. After a loss of 6.0 per cent of jobs in the early 1990s, recovery from spring 1993 in the workforce in employment was slow, reaching only 2.5 per cent in total by end-1995. With reductions in that period in services such as transport and banking, the number of full-time employees continued to fall, and the number of economically active adults increased, allowing women's and insecure forms of work to gain in relative importance.

222

THE 'JOBS GAP'

The overall monetary future of the EU and its level of unemployment have been displacing from discussion social policy for the protection of workers' rights (Gold, 1993) and particularly those of women workers. Although the growth of GDP in the EU resumed in mid-1993, the impact on numbers out of work was minimal. In the stagnant job market of 1995, when average EU unemployment remained at 11.0 per cent, the number of jobless was expected to fall only marginally before the middle of 1997 (CEC, 1995b). An increase in jobs of about 1 per cent per annum would compensate little for the job losses suffered over the preceding three years, following the precedent of the mid-1980s when it took three years after unemployment had reached its peak for the rate to begin to fall significantly.

With a little variation between countries, growth in GDP has to exceed 2 per cent per annum to have any impact on employment levels:

It will probably remain inescapable that the elimination of European unemployment is a very difficult challenge. The work reviewed here suggests strongly that in Europe the mechanism through which unemployment could be self-correcting is weak and slow. A gradual elimination would call for a prolonged period during which output grows faster than productivity – a situation not witnessed over the past thirty or forty years.

(Dreze and Bean, 1990: 60)

The CEC (1995b) agree that a high rate of GDP growth is not a sufficient condition for maintaining a high level of employment, but see it as a prerequisite. In the second half of the 1980s, no member state increased employment by more than 1 per cent a year without average production growth of around 3 per cent or more. Job growth has been closely related to growth of GDP, but the underlying growth of productivity has been the problem. Over the 1980s the growth of labour productivity (defined as GDP per person employed) stood remarkably constant at just under 2 per cent per year, both in periods of slow growth and in the more dynamic second part of the 1980s. What is more, the same rate held during the ensuing recession – that is to say that employers succeeded in running down their workforces to match demand.

Recession had a severe impact on employment in most member states (Table 11.3). The numbers employed in the EU as a whole declined by 4 per cent in the three years 1991 to 1994, twice as much as any comparable period since the war. As a result, 6 million jobs – some 60 per cent of the 10 million or so net increase from the earlier period, 1985 to 1991 – were effectively lost over this period. Given the growth of working-age population and trends in rates of participation, the EU's labour force would otherwise have been expected to expand by some 4 million. The EU is some 10 million jobs short.

Table 11.3 Percentage change in total employment, E15, 1975–94

	1975–85	1985–91	1991–94
E15	+2.3	+7.6	−4.0
Denmark	+8.6	−0.2	−1.2
Finland	+5.4	−4.4	−13.5
Sweden	+6.4	+3.7	−6.5
Austria	+9.9	+7.7	+1.6
Belgium	−2.8	+5.9	−2.5
France	+3.6	+4.1	−1.9
Germany[1]	+3.5	+10.2	−4.5
Ireland	+1.7	+5.1	+3.8
Luxembourg	+2.5	+22.4	+6.6
The Netherlands	+10.4	+25.3	+4.5
United Kingdom	−0.6	+5.8	−3.4
Greece	+10.0	+1.2	+3.7
Italy	+4.8	+4.1	−3.5
Portugal	+12.7	+11.2	−4.1
Spain	−12.2	+17.9	−6.2

Note: 1 1994 data includes the 'new Lander' of Eastern Germany
Source: CEC, 1995b

The table finds the most rapid rates of job loss in Finland and Sweden, both affected by reductions of welfare state spending; Finland has also lost trade with Russia. Other leading areas of job loss were in Italy, Spain and the UK, all notable for job increase in the previous period of expansion, 1985 to 1991. Countries that had actually achieved job growth in that period all retained GDP growth of 2 per cent per year. In the previous period, the three member states achieving the largest gains in employment – The Netherlands, Luxembourg and Spain – were also the three experiencing the smallest gains in output per person employed – i.e. the most labour-intensive growth. Other data supports the view that lower productivity growth, not surprisingly, does provide a better achievement in terms of numbers employed. In the USA, employment-intensive growth has been concentrated in non-manufacturing sectors where high productivity is less of a priority because the output does not generally enter international trade. The poor quality of some US jobs (Chapter 5) and their continuing low pay must discourage the EU from seeking to adopt this solution.

The boldest EU programmes have sought no more than a halving of unemployment, whether in the Union as a whole or in individual countries. In the UK a study for the Rowntree Foundation (Meadows, 1996) found that economic growth would not make a substantial impact on unemployment, nor would existing training and job-seeking programmes, or attempts to increase the numbers in work through social security changes; as a result, notably, households with no wage earners in the UK would rise to 20 per cent of the total, up from 5 per cent fifteen years ago.

PRESENT EU POLICIES FOR STRUCTURAL ADJUSTMENT

If asked what the CEC was doing about employment problems, officials might respond with reference to the Structural Funds. Although these are the most relevant tools considered so far, this could lead to several misunderstandings. First, Structural Funds relate only to specific problems arising in particular places and sectors. Second, they are subject to severe financial restrictions, with which I shall begin.

Begg and Mayes (1993) reported to the European Parliament that, compared with member states, the CEC had little room for manoeuvre in effecting any redistribution. The overall budget stood at only 1.2 per cent of the EU's total GDP, and member states proved reluctant to raise this proportion. A large share of this amount, over half, goes on agriculture and this is distributed not in relation to poverty or deprivation on the land but in relation to output. Overall, as Franzmeyer *et al.* (1991) show, the impact of the CEC budget on less favoured regions is in fact neutral, achieving little correction to trends in unemployment or employment (Chapter 3, above). Ireland and Greece were net beneficiaries, but so too was one of the richest member states, Denmark, while Spain saw little net benefit. The Community's Structural Funds comprise the following:

- **European Regional Development Fund (ERDF),** supporting investment in infrastructure, assistance to small and medium-sized firms, and tourism and environmental projects.
- **European Social Fund (ESF),** supporting training, retraining and vocational guidance with particular emphasis on young people and the long-term unemployed.
- **European Agricultural Guidance and Guarantee Fund (EAGGF),** supporting measures to speed up the adjustment of agricultural structures, with a view to the reform of the Common Agricultural Policy.
- **Financial Instrument for Fisheries Guidance (FIFG),** to speed up adjustment of Fisheries Policies.

Under these arrangements, the Structural Funds are allocated to seven priority Objectives against which areas are classified (1994–9):

Objective 1: development and structural adjustment of the regions whose development is lagging behind – ERDF, ESF, EAGGF.
Objective 2: converting areas affected by industrial decline – ERDF, ESF.
Objective 3: dealing with long-term unemployment, and facilitating the

integration of those threatened with exclusion from the labour market – ESF.

Objective 4: facilitating workers' adaptation to industrial changes and changes in production systems – ESF.

Objective 5a: speeding up adjustment of agricultural structures as part of CAP reform – EAGGF, FIFG.

Objective 5b: facilitating the development and structural adjustment of agricultural areas – ERDF, ESF, EAGGF.

Objective 6: to provide for development of regions with an extremely low population density (Finland and Sweden).

(Objectives 3 and 4 include new tasks dating from the Maastricht Treaty of 1991, and will be considered among the implications of Community-wide policy, following immediate attention to spatial policies under the European Regional Development Fund [ERDF]).

Traditional spatial policies

The European Regional Development Fund provides co-financing for regional development projects in assisted areas identified by national governments. As such, the European Commission essentially showed support for areas that were 'peripheral' to member countries and sometimes to the combined areas of member states. Thus, when the UK joined, areas of Scotland that were already recognised by the UK government as Development Areas became eligible for infrastructure grants along with equivalent areas of Brittany or Corsica as recognised by the French government, or of Southern Italy as recognised by Rome. However, for a long while there was confusion as to whether any evaluation of results was taking place, and whether, having acquired particular grants for their regions, member states were not simply 'pocketing the money' rather than implementing any more expenditure than was already taking place. This issue of 'additionality' still provides many technical and political problems (Bachtler and Michie, 1995); the UK still tends to see the funds as partial reimbursement of the country's net contribution to the EU budget, while others identify the member states as the real winners from the whole process, as they may have retained assisted areas for the very purpose of attracting European money.

Reforms of 1988 had improved the scope, scale and rigour of the Structural Funds, which were then related to specific structural problems embodied in the five objectives, all to be assessed in periodic reports such as CEC (1994b). These reports were, however, cautious about attributing socio-economic effects and trends directly to Structural Fund expenditure. Indeed, they counted for little against the effect, for instance, of population trends in raising unemployment in Objective 1 areas, but could be associated with reduced unemployment in places classified under Objective 2 (Chapter 3).

Regional disparities in unemployment fell in the late 1980s but grew again by 1994, especially in much of France, Spain and Greece, together with major cities such as Brussels and Berlin. The Maastricht Treaty of 1991 was seen to expose places to the threat of increased competition at the hands of the EU itself, and led to a review of the effectiveness of the Structural Funds, followed by a fresh designation of assisted areas (Figure 11.1).

The history of regional policy in the 1980s in the nations of E9 had been one of declining expenditure (1980–90) as a percentage of GDP in all except Italy, and at constant prices in five countries (CEC, 1994b). There was a view that assistance should be focused on the most deserving projects and areas, reflecting a trend towards greater selectivity and discretion in the use of incentives. In any case, the amount of industrial investment available for dispersal from prosperous areas was more limited in a period of reduced growth (Bachtler and Michie, 1993). The UK reduced the coverage of its assisted areas from 49.5 per cent of the national population in 1980 to 36.8 for the period 1984 to 1994, while the average for E9 shrank from 37.9 per cent to 35.5 in 1986 and 33.4 in 1992. That, however, leaves out the new members, the accession of which had a profound effect on the balance of need and the political calculations of Brussels. With Greece, Italy and Spain counted in E9, the proportion of population of the EU in assisted areas in 1986 was 41.0 per cent.

A major result of consideration at the beginning of the 1990s was that weaker Southern European countries as a whole were recognised as needing some compensation to assist them in meeting the Single European Market: thus Portugal, Spain and Greece, as well as Ireland, were designated to receive payments from a Cohesion Fund. As regards actual assisted areas, the 1993 review increased the above figure of 41.0 per cent to 51.6 per cent (Figure 11.1) and this is the proportion counted in Objective 1, 2 and 5b areas until the next change. This was the addition of Objective 6 in recognition of special problems in the next new members, Sweden and Finland. The whole of East Germany with its serious problems had already been adopted as a high priority area, but heavy unemployment in the industrial economy as a whole led to the addition also to Objective 1 of new areas of France (Nord-Pas-de-Calais), Belgium (Hainaut), The Netherlands (Flevoland) and the UK (the Highlands and Merseyside).

Figure 11.1 therefore represents a more complex undertaking than at any time in the history of European regional policy, of importance for local authority councillors and planners in virtually all corners of the EU. However, it is very important to note that a fundamental review of the Structural Funds is beginning in 1997 prior to the expiry of the map in Figure 11.1 in 1999. It is thought that this might involve reduction and greater targeting of funds in the face of many criticisms. Harrop (1996) questions whether the EU is capable of tackling unemployment effectively, especially through the Structural Funds, citing the failure to reduce unemployment in Objective 1

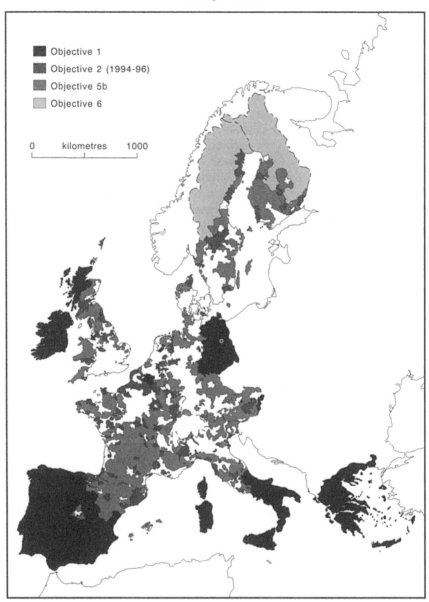

Figure 11.1 Eligibility for EU Structural Funds, 1994–9, E15
Source: various EU documents

areas, and the reduced support from national government budgets. One of
the biggest existing problems is that existing:

regional development strategies are often only loosely related to the problems of the regions they are meant to address. In many cases they have been drafted primarily as a vehicle for 'drawing down' expenditure rather than as a coherent strategy.

(Bachtler and Michie, 1995: 747)

Many inconsistencies remain between countries and regions in commitment and approach to evaluation. This is not only in data and definitions but also in assessing multipliers or unproductive investment (for instance where one tourism project displaces trade from an existing one, or where a project would have taken place without regional financial assistance).

In one set of forecasts, Cambridge Econometrics (1995) reported that many of the manufacturing cities that suffered most in the recession of the early 1990s were among those leading recovery. However, their forecasts for 1993–9 showed marked divergence between areas expanding their Gross Value Added by 3.6 to 4.1 per cent per year, in northern Italy, large parts of The Netherlands and Berlin, and the areas of slowest growth, of 1.2 to 1.7 per cent per year, concentrated in Greece and rural France. 'The regional problem', understood as such, is set to continue, whether or not it is a focus of policy and investment.

Regional policy belongs to a mind-set now to some extent overtaken by neo-liberal policies of competition in the 1990s. CEC officials have found that countries bid against each other to attract major industrial projects, leading to a spiral of countries offering higher and higher subsidies to gain jobs in areas of structural weakness. Poor countries may simply lack sufficient funds to join in the contest. On the other hand, fears that state aid to problem regions was leading to severe distortions of competition have prompted a thorough examination of regional aid rules (*Financial Times*, 28 February 1996: 3). The worry was that the combination of European, national and local schemes in disadvantaged areas such as Eastern Germany and Southern Italy was leading to such an escalation of aid, especially in large capital-intensive projects, as to undermine the Single Market.

The incorporation of Portugal, Spain and Greece, which changed EU policy so greatly, is certain to be followed at some stage by the incorporation of relatively poor countries from Eastern Europe, as agreed in principle in mid-1993. Change in the CAP was seen as an essential prerequisite, and

reform of the structural funds governing regional and social spending is also seen as essential, because without it, all the available cash would be transferred to eastern Europe, cutting off billions of Ecus in aid to countries such as Greece, Ireland and Portugal, as well as the UK's regions of high unemployment. Whatever happens, they face funding cuts.

(Barber *et al.*, 1995: 1)

COHESION OF THE EU AS A WHOLE

As I write, German people are protesting publicly about spending cuts deemed necessary by the government to meet the criteria for the use of a common European currency in 1999. The stages of union from 1989 to 1999 are seen as strengthening the global competitiveness of the EU, but also as imposing strain on many regions. If the CEC provide extra funds to deal with problems of 'cohesion', are these just side-payments extended in exchange for other policies, assisting the main process of letting capitalism rip? 'Structural adjustment is now affecting a much wider range of regions throughout Europe' (Bachtler and Michie, 1994: 789). I want here to stress more than ever that employment and social problems in the 1990s have outrun the limits, wide as they are, of assisted areas and regional policies.

Hadjimichalis and Sadler (1995) argue that marginality resulting from social and spatial processes has been accentuated at one and the same time across the EU. 'What is important in the last decade or so, however, is the precise way in which new forms of differentiation have been enabled to emerge, and old ones have taken on an even sharper edge' (Hadjimichalis and Sadler, 1995: 6). Beneath the veneer of consensus, they argue, lies an under current of tension; the same policies that were creating a competitive or 'neo-liberal' thrust to the agenda of integration were also ignoring growing inequalities among gender, age, ethnic, religious and employment groups.

Individual countries have come to see spatial problems as merely one expression of deeper structural difficulties, and so must the EU. The Single European Act, the Maastricht Treaty and progress toward a European Currency (scheduled at the time of writing for 1999; Figure 11.2) are all seen as having far-reaching effects on great areas of the Community, on member states as a whole and on policies that are more vulnerable to particular financial changes, above all the cutting of public spending. To enter the common currency, countries will be expected to have similar defined levels of exchange rates, inflation and deficits in public spending. This will 'entail restrictions on macro-economic policy that are more onerous for the less-prosperous countries of the EC' (Begg and Mayes, 1993: 151), in turn having a disproportionate effect on poorer people, cities and regions. Governments will have little latitude for policy of any kind, even in the richest countries and areas. Amin and Tomaney (1995) are quite clear that plans for economic and monetary union are likely further to widen regional disparities.

> The removal of exchange rates as an instrument of economic policy deprives nation-states of the flexibility to meet external shocks that necessarily affect each member state, and the regions within it, differently ... the most serious threat is to Less Favoured Regions in member states with weak fiscal capacity and which are running large deficits.
>
> (Amin and Tomaney, 1995: 177).

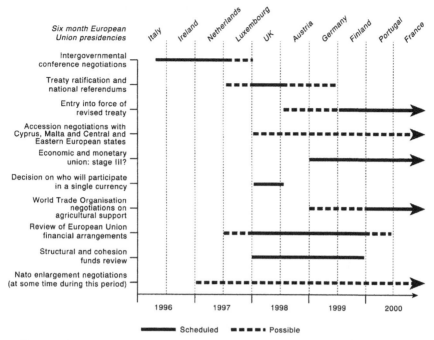

Figure 11.2 The EU inter-governmental conference; intended schedule at time of
writing
Source: *Financial Times* 29 March 1996: 2

Grave employment problems remain in London, Paris or Brussels, but
nonetheless the CEC itself rightly accepts that the Single European Market
might weaken Objective 1 and 2 regions relative to core areas of the EU.
This would not be self-correcting: on the one hand, the central city regions
of the EU would benefit all the more from their concentrations of skills in
technology and R&D; on the other, the free market across the EU would
expose Less Favoured Regions to market forces working in favour of the
most powerful or most competitive firms in the national and international
economy, encouraged to merge and to rationalise their branches under EU
competition policy (Chapter 4). It was feared that many firms, especially in
the Less Favoured Regions, would collapse as demand was met by transnational companies reaping the benefits of scale and the organizational
power they possess. These processes may only repeat the story of Southern
Italy and East Germany (Chapters 1 and 2).

The EU were aware that acceleration of economic integration in Europe
would have such effects as to endanger convergence and merit special policies, without which further integration might be endangered. In taking the
term 'cohesion' as a watchword for keeping everybody on board, the CEC

might just be concealing the effects of freer trade by European TNCs. Despite the arguments that regional bodies and the EU were gaining relative power in a 'Europe of the Regions', the member states chose under the Maastricht Treaty to establish a Cohesion Fund to assist transport and environmental projects in the four member states whose GDP per head was less than 90 per cent of the Community average, i.e. Greece, Portugal, Ireland and Spain.

Such action has both financial and political limits. Many northern members will resent any real expansion in these funds, and, if their own levels of unemployment continue high, will continue to attract investment which in the past might have decentralised to Southern Europe.

THE FORWARD VIEW

The unfolding of the 1990s only deepened the crisis of confidence in the EU, with even Germany losing some of its hopes of and credentials for European Monetary Union (EMU). Unemployment, spending cuts and the insecurity prompted by 'reform' of welfare states all weakened national governments, as illustrated by French public sector strikes of 1995, prompted by financial measures needed to prepare for EMU. The OECD increasingly argued that the move towards a single currency in Europe could increase unemployment in the absence of labour market reform, while Coates and Holland (1995) argued that reducing debts and deficits to required levels could add 10 million to the number unemployed.

Demographic, sectoral, cyclical and policy factors will interact in determining the precise future of the 'jobs gap'. Job loss after 1990 significantly staunched the whole growth of an active labour force across the EU, and in a number of member states resulted in an absolute reduction in the economically active. Experience of the 1980s strongly suggests that a period of job creation will bring more people back on to the labour market, requiring a greater increase of jobs than is at all probable to make any effect on unemployment totals. Different factors affect different age groups. How many fewer young people would enter further education if more jobs were open to them? It is likely that the number of women in less prosperous EU countries in paid work will increase, adding to pressure for job creation; so might a return to work of men who have dropped out of the labour force.

Fears of a jobless recovery were slightly allayed by a forecast of unemployment falling in the EU to 9 per cent by 1997 (CEC, 1995b). The recession of the early 1990s accentuated the shift away from manual jobs (skilled as well as unskilled) towards non-manual work, especially highly skilled jobs that require much education and training, like the 'symbolic-analytic services', the only jobs found to be increasing in Reich's (1991) analysis of the USA. Workers with relatively few qualifications have both suffered job losses directly (especially those employed in industry) and, as competition for scarce jobs has intensified, have found it difficult to obtain alternative employment. 'The same tendencies are evident for both men and women . . . [and] are also apparent in most member states' (CEC, 1995b: 62).

Unsurprisingly the UK yields similar forecasts. The Institute for Employment Research (1994) added, however, that technological and other changes were starting to cause job loss in lower level non-manual jobs, especially notable already in banking and insurance. Business services and tourism were showing much of the form that selected them for earlier chapters. Other parts of the service sector to be showing job loss already in 1995 included retailing, public administration, electricity and gas, as well as many parts of transport and communication where decline was much longer established. All in all the UK was forecast to gain jobs in services, women's work and self-employment, suggesting an unemployment rate in 2001 of 8 per cent, with little change in the north–south divide.

FUTURE EMPLOYMENT POLICY?

For the EU to suffer unemployment of more than say 5 per cent for decade after decade would be a severe indictment of European civilisation and government. Jobs require greater priority than they have enjoyed. No one knows how low unemployment can fall without causing an acceleration of inflation – this rate is generally agreed to have increased compared with previous decades for the EU – but the years 1994 to 1996 saw increased recognition that the EU rate of unemployment must be reduced.

Prior to that date, 'employment policy' could be seen as a matter of the Charter of Fundamental Social Rights for Workers adopted in 1989 by eleven member states, excluding the UK. The Charter was concerned with improved living and working conditions, equal treatment of men and women, health and safety at work and freedom of association, etc. In the 1990s, the debate has moved on with, for instance, unprecedented levels of unemployment in Germany causing a re-evaluation of the costs of labour and employment protection relative to job creation.

By 1996, there had been a shift along this axis of debate away from the legal imposition of minimum standards, on gender equality for example, toward an emphasis on economic growth, sometimes incorporating the 'Green' view that environmental policies could create jobs. The views that I put forward here are a mixture of conventional wisdom on the one hand, and on the other an awareness from low income countries of the opportunities for job creation in local, labour-intensive services.

Investment in vocational training

Many observers, including the UK Labour Party, give this high priority. The shift in demand from unskilled to skilled workers must be met, to improve opportunities for the labour force, to seek to redress youth unemployment and to upgrade the contribution of women to the economy. This last is essential since women account for the whole growth in the workforce (CECb, 1995). However, in closing a major review of EU policies, Harrop (1996: 157) notes that 'at the present time, 1995, what is mainly visible is a better educated dole queue'.

Increasing the employment intensity of growth

If there are differences between countries in the growth needed in GDP to create a certain increase in employment, it might be that these demonstrate a greater job yield in one sector than another. The search for productivity gains in traded manufacturing goods would normally militate against a labour-intensive policy in this sector, but there appear to be countries in which low growth in productivity in services has contributed to a better record in job creation. Although not ideal in the long term, 'the effect on production costs in the traded-output sectors of higher than minimum costs in other parts of the economy will in many cases be very small' (CEC, 1995b: 15). The CEC to some extent see the growth of *self-employment* as limited, as failure of many firms offsets the successes that are reported. The cultivation of *part-time employment* can significantly stave off the reduction of women's jobs in recession, and merits the addition of pension rights for part-timers, as ensured by the EU in 1995.

Job-sharing is increasingly recognised as a viable means of achieving several goals and is perhaps the most widely understood change towards a better spread of work. Job-sharing receives the endorsement of CEC (1995b), but they calculate that results are meagre.

A shorter working week would seem an ideal answer to the problem of unemployment. In a number of member states, legislation has been introduced in this area, though it has tended to take the form of reducing the maximum number of hours worked per day or per week primarily for health and safety reasons rather than to reduce the number of people at work. Costs and benefits are seen to vary between sectors, and actual progress has been remarkably slow. During the rapid growth from 1985 to 1990, average hours per week dropped from forty to thirty-nine (adding presumably 2.5 per cent to the number of people with jobs). In the subsequent recession from 1990 to 1994, the rate of reduction in hours was reduced and was confined to services.

Reducing non-wage labour costs 'to stimulate job creation' is one of the newer and more contentious areas of neo-liberal discourse. Graphs are increasingly put round the world showing national differences in taxes paid by employers on top of the wages they pay to workers (which may also be taxed). In 1993 in the EU, the overall government charge in compulsory contributions like national insurance and pension contributions from employers for a single worker with half the average earnings was as high as 40 per cent of labour costs in five member states (although much lower for workers with families). In the USA, the equivalent figure was about 25 per cent, and the difference is seen in part as explaining the 'success' of employment in the US. The European preference for a better welfare state than in the USA is clear (Chapter 5), but the scale of these 'taxes on employment' is increasingly being questioned by neo-liberals, and in 1996 acted on, for instance by the German government.

The effectiveness of labour market policies has many other aspects, including the role of government placement, mobility and training schemes. Much of this is, however, frustrated by the scale of unemployment itself. Job creation may not help the position of the long-term unemployed because of their unfamiliarity with work.

Individual countries are adopting different mixtures of these policies for their domestic problems, for instance France in February 1996, including new tax concessions for deprived inner city zones and certain rural areas, and Belgium in April 1996. The explicit aim still rarely exceeds that of halving unemployment within a definable period. In December 1995, the EU announced the aim of halving unemployment by the year 2000 but many Finance Ministers distanced themselves from this, expressing reservations about creating a net increase of 11 million jobs in five years. That figure itself represented a reduction from the previous objective of 15 million, and half of the shift would have to arise from reductions in working hours, which we have described above as being surprisingly intractable.

I have shown the continued scope for deindustrialisation through productivity gains in manufacturing, and the confinement of other job gains mainly to tourism and certain business services. In this context, the addition of all the types of measures above would still not lead us toward an optimistic conclusion. The picture can be altered only by a really long sustained boom, and a successful combination of EU and national policies.

RADICAL NEW POLICIES – THE LOCAL AND THE ENVIRONMENTAL

However, the wheel of fashion is producing possible *contributions* to the problem in Brussels thinking. Following to some extent what I have seen in some low income countries, I am convinced that, far removed from Brussels, voluntary initiatives can be valuable for local people at least at the scale of rural areas. These might be in establishing services, environmental improvements and other work projects that generate jobs (see below). In proportion, however, to the scale of urban and national economies, it is not so clear that these politically novel approaches can subsist without wider change in financial and political structures. Brussels has, however, shown remarkable interest, reserving a proportion of the Structural Funds under the 1994 reforms for local Community Initiatives and starting a study (CEC, 1995a) of all the employment possibilities from such schemes.

Services in general can be part of the economic base of an area (Townsend, 1991a); it is sensible and at the top of my agenda to encourage services that draw in income to an area, for instance appropriate retailing or tourism. The thrust of a number of new arguments is that direct cultivation of purely or mainly local service jobs is legitimate and beneficial. The British Labour Party has long argued that government and local authority services can be expanded to make large inroads into the total numbers unemployed, and now the UK 'Full Employment Forum' (Berry, 1995) seeks to generate *The First Million Jobs* in housing, education, health, personal social services and environmental jobs.

What is new is the EU recognition (in CEC, 1995a) that there were significant and unmet needs in the service sector that could best be met on a 'bottom-up' basis. We can document that there already is such growth in jobs. For instance, GB employees in 'social work' increased by 2.4 per cent in 1994 alone; 'personal and collective services' have been a major area of European job growth even during the recession 'and offer the potential for even more significant expansion in the years ahead' (CECa, 1995: 21). The new aim here is to establish a framework for meeting local needs that arise from either improving standards of living or changing patterns of behaviour, and which have so far been less than adequately served by both public and private sectors.

This could add 0.1 to 0.3 per cent a year to the total of EU jobs – or some 140–400 thousand per year (CEC, 1995a). The fields were:

- Services at home
- Child care
- Assistance to young people needing reintegration into society
- Security
- Local public transport
- Redevelopment of public urban areas
- Local shops
- Tourism
- Audiovisual techniques
- Cultural heritage
- Local cultural development
- Waste management
- Water management
- Protection and maintenance of green areas
- Pollution regulations and controls, and related equipment

The EU agrees that the rate of job creation would *benefit*, rather than suffer, from a higher level of environmental protection in Europe. Deindustrialisation would tend to eliminate older sources of pollution from heavy industry, while environmental programmes are seen as creating jobs. Almost 1.5 million people are thought to be employed in connection with environmental protection. Most were in sewage, water disposal, water supply and recycling, but instrument engineering was also profiting from the imposition of controls and was one of the few expanding parts of the engineering sector. In the wider world, as in the EU, environmental and labour standards need international enforcement.

THE END AND THE BEGINNING

On what we know, I believe that radical new policies in caring for people and environment have the most to offer grassroots activists in the creation of jobs. Agreed shorter working hours should be of value at a national and EU level, and much more thought and effort in this field are urgent. At all levels, what can be achieved will depend on the alliances that we can construct (Giddens, 1996). The medium-term fate of many local areas in the EU will depend on the success of individuals, NGOs, entrepreneurs and politicians not only in identifying opportunities and potential alliances but in evaluating them in their national, EU and global context: a case for geographical integration of interests in space. Making a livelihood in Europe matters to us all, in academic, planning and everyday life.

> In a world where the amount of available work may shrink substantially over the coming twenty years, the distribution of work holds the key to overall social integration.
>
> (Giddens, 1996: 161)

BIBLIOGRAPHY

Ahnstrom, L. (1990) *Economic Growth, Stagnation and the Working Population of Western Europe*, London: Belhaven.

Akehurst, G. (1992) 'European Community tourism policy', in P. Johnson and B. Thomas (eds) *Perspectives on Tourism Policy*, London: Mansell.

Alexander, N. (1995) 'UK retail expansion in North America and Europe', *Journal of Retail and Consumer Studies* 10, 2: 75–81.

Allen, J. (1992) 'Services and the UK space economy: regionalization and economic dislocation', *Transactions, Institute of British Geographers*, New Series 17, 3: 292–305.

Allen, J. and Pryke, M. (1994) 'The production of service space', *Environment and Planning D, Society and Space* 12: 453–77.

Amin, A. and Thrift, N. (1994) 'Living in the Global', in A. Amin and N. Thrift (eds) *Globalisation, Institutions, and Regional Development in Europe*, Oxford: Oxford University Press.

Amin, A. and Tomaney, J. (1995) 'The challenge of cohesion', in A. Amin and J. Tomaney (eds) *Behind the Myth of European Union*, London: Routledge.

Anderton, B. and Mayhew, K. (1994) 'A comparative analysis of the UK labour market', in R. Barrell (ed.) *The UK Labour Market*, Cambridge: Cambridge University Press.

Artobolevsky, S.S. (1996) *Regional Policy in Europe*, London: Jessica Kingsley.

Ashworth, G.J. and Bergsma, J. (1987) 'Policy for tourism: recent changes in the Netherlands', *Tidjschrift voor Economische en Sociale Geografie* 78, 2: 162–75.

Ashworth, G.J. and Tunbridge, J.E. (1990) *The Tourist-Historic City*, London: Belhaven.

Ashworth, G. and Voogd, M. (1988) 'Marketing the city: concepts, processes and Dutch applications', *Town Planning Review* 59, 1: 65–79.

Atkinson, J. (1985) 'Flexibility: planning for an uncertain future', *Manpower Policy and Practice* 1, summer: 26–9.

—— (1988) 'Recent changes in the labour market structure in the UK', in W. Buitelaar (ed.) *Technology and Work: Labour Studies in England, Germany and the Netherlands*, Aldershot: Avebury.

Atkinson, J. and Meager, N. (1986) *Changing Work Patterns: How Companies achieve Flexibility to meet New Needs*, London: National Economic Development Office.

Bachtler, J. and Michie, R. (1993) 'The restructuring of regional policy in the European Community', *Regional Studies*, 27, 8: 719–25.

—— (1994) 'Strengthening economic and social cohesion? The revision of the Structural Funds', *Regional Studies*, 28, 8: 789–96.

—— (1995) 'A new era in EU Regional Policy evaluation? The appraisal of the structural Funds', *Regional Studies*, 29, 8: 745–51.

Bade, F.J. and Kunzman, K.R. (1991) 'Deindustrialisation and regional development in the Federal Republic of Germany', in L. Rodwin and H. Sazanami (eds) *Industrial Change and Regional Economic Transformation*, London: HarperCollins.

Bagguley, P. (1990) 'Gender and labour flexibility in hotels and catering', *The Service Industries Journal* 10, 4: 737–47.

Bagguley, P., Mark-Lawson, J., Shapiro, D., Urry, J., Walby, S., and Warde, A. (1990) *Restructuring: Place, Class and Gender*, London: Sage.

Baldwin, R. (1992) 'Measurable dynamic gains from trade', *Journal of Political Economy* 100: 162–74.

Ball, R. (1989) 'Some aspects of tourism, seasonality and local labour markets', *Area* 21, 1: 35–45.

Balls, E. (1994) 'No more jobs for the boys', in J. Michie and J. Grieve-Smith, *Unemployment in Europe*, London: Academic Press.

Barber, L., Peel, Q. and Peston, R. (1995) 'Santer plans reform of regional aid funds', *Financial Times,* 18 May: 1995, 1

Baret, C. and Bertrand, O. (1992) *Training in the Retail Trade: Report for the FORCE Programme*, Berlin: CEDEFOP.

Barlett, D.L. and Steele, J.B. (1992) *America: What Went Wrong?*, Kansas City: Andrews McMeel.

Barras, R. (1983) *Growth and Change in the UK Service Sector*, London: Technical Change Centre.

Bateman, M. and Burtenshaw, D. (1979) 'The social effects of office decentralization', in P.W. Daniels (ed.) *Spatial Patterns of Office Growth and Location*, Chichester: John Wiley.

Baty, B. and Richards, S. (1991) 'Results from the Leisure Day Visits Survey 1988–89', *Employment Gazette* 99, 5: 257–68.

Beatson, M. (1995) 'Progress towards a flexible labour market', *Employment Gazette* 103, 2: 55–66.

Begg, I. and Mayes, D. (1993) 'Cohesion, convergence and economic and monetary union in Europe', *Regional Studies* 27, 2: 149–55.

Bell, D. (1974) *The Coming of Post-Industrial Society*, London: Heinemann.

Bennison, D. and Boutsouki, C. (1995) 'Greek retailing in transition', *International Journal of Retail & Distribution Management* 23, 1: 24–31.

Berry, R. (1995) 'The case for increased public spending', in R. Berry (ed.)*Economic Policies for Full Employment and the Welfare State*, London, PO Box 188, SW1A 0SGA: Campaign to Defend the Welfare State/ Full Employment Forum pamphlet.

Beyers, W.B. (1991) 'Trends in producer services in the USA: the last decade', in P.W. Daniels (ed.) *Services and Metropolitan Development*, London: Routledge.

Beyers, W.B., Tofflemire, H.A., Stranahan, H.A. and Johnson, E.K. (1986) *The Service Economy: Understanding the Growth of Producer Services in the Central Puget Sound Region*, Seattle: Central Puget Sound Economic Development District.

Bhalla, A.S. (1973) 'The role of services in employment expansion', in R. Jolly, E. de Kadt, H. Singer and F. Wilson (eds) *Third World Employment*, Harmondsworth: Penguin.

Bianchi, S. M. and Spain, D. (1986) *American Women in Transition*, New York: Russell Sage.

Black, S. and Conway, E. (1995) 'Community-led Rural Development Policies in the Highlands and Islands: The European Community's LEADER Programme', *Local Economy* 10, 3: 229–45.

Blackaby, F. (ed.) (1979) *Deindustrialisation*, London: Heinemann.

Bodrova V. and Anker, R. (1985) *Working Women in Socialist Countries: the Fertility Connection*, Geneva: International Labour Organisation.

Boor, M. (1980) 'Relationships between unemployment rates and suicide rates in eight countries', *Psychological Reports* 47, 4: 1095–101.

Boulianne, L.M. and Decoutere, S. (1994), 'La structuration spatiale des activités de services aux enterprises: la suisse romande et la Region metropolitaine Montrealaise', paper presented at the Fourth RESER Conference, Lyon, France, September.

Bowley, G. (1996) 'Economists ponder shorter dole queues', *Financial Times* 12 June: 14.

Braverman, H. (1974) *Labor and Monopoly Capital: The Degradation of Work in the Twentieth Century*, New York: Monthly Review Press.

Breathnach, P., Henry, M., Drea, S. and O' Flaherty, M. (1994), 'Gender in Irish tourism employment', in V. Kinnaird and D. Hall (eds) *Tourism: A Gender Analysis*, Chichester: John Wiley

Brewster, C. (1995) 'You've got to go with the flow', *The Guardian, Careers* 22 April: 2–3.

Bridges, W. (1995) *Jobshift: How to Prosper in a Workplace without Jobs*, London: Nicholas Brealey.

Brittan, S. (1993) 'The two-way switch in world economy', *Financial Times* 9 December: 22.

Britton, S. (1990) 'The role of services in production', *Progress in Human Geography* 14: 529–46.

Brusco, S. (1986) 'Small firms and industrial districts', in D. Keeble and E. Wever (eds) *New Firms and Regional Development in Europe*, London: Croom Helm.

Bull, P.J. and Church, A. (1994) 'The geography of employment change in the hotel and catering industry of Great Britain in the 1980s: a subregional perspective', *Regional Studies* 28, 1: 13–25.

Burt, S.L. and Dawson, J.A. (1990) 'From small shop to hypermarket: the dynamics of retailing', in D. Pinder (ed.) *Western Europe: Challenge and Change*, London: Belhaven.

Burton, R.C.J. (1994) 'Geographical patterns of tourism in Europe', *Progress in Tourism and Recreation Hospitality Management* 5, 1: 3–25.

Business Strategies (1996) *Labour Market Flexibility and Financial Services*, London: Business Strategies.

Caire, G. (1989) 'Atypical wage employment in France', in G. Rogers and J. Rodgers (1989) *Precarious Jobs in Labour Market Regulation: The Growth of Atypical Employment in Western Europe*, Geneva: International Labour Organisation.

Cambridge Econometrics (1993) *European Regional Prospects*, Cambridge: Cambridge Econometrics.

—— (1995) *European Regional Prospects*, Cambridge: Cambridge Econometrics.

Campagne, P. (1990) 'Three agricultural regions of France: three types of pluriactivity', *Journal of Rural Studies* 6, 5: 415– 22.

Campbell, B. (1993) *Goliath: Britain's Dangerous Places*, London: Methuen.

Campbell, M. and Daly, M. (1992) 'Self-employment into the 1990s', *Employment Gazette*, 100, 6: 269–92.

CEC, Commission of the European Communities (1987) *Europeans and their Holidays*, Brussels: Commission of the European Communities.

—— (1991a) *Fourth Periodic Report on the Social and Economic Situation and Development of the Regions of the Community*, Brussels: Commission of the European Communities.

—— (1991b) *Employment in Europe, 1991*, Brussels: Commission of the European Communities.

—— (1993a) *Portrait of the Regions*, 3 vols Brussels: Commission of the European Communities.

—— (1993b) *The Evolution in Holiday Travel Facilities and the Flow of Tourism inside and outside the European Community*, Brussels: Commission of the European Communities.

—— (1993c) *Urban Regeneration and Industrial Change*, Brussels, Commission of the European Communities.

—— (1993d) *Impact of the Completion of the Internal Market on the Tourism Sector*, Brussels: Commission of the European Communities.

—— (1994a) *Employment in Europe, 1994*, Luxembourg: European Communities.

—— (1994b) *Fifth Periodic Report on the Social and Economic Situation and Development of the Regions of the Community*, Brussels: Commission of the European Communities.

—— (1994c) *Europe 2000+: Cooperation for European Territorial Development*, Brussels: Commission of the European Communities.

—— (1995a) *Local Development and Employment Initiatives*, Brussels: Commission of the European Communities.

—— (1995b) *Employment in Europe, 1995*, Luxembourg: European Communities.

—— (1995c) *Regional Development in Rural Areas, 1994–99*, Brussels, Commission of the European Communities.

—— (1995d) *The Craft Industry and Small Enterprises*, Brussels: Commission of the European Communities.

CEC, Eurostat (1991) *Eurostat: Tourism, 1988–9, Annual Statistics*, Brussels, Commission of the European Communities.

—— (1993a) *Labour Force Survey, Results, 1991*, Luxembourg, European Communities.

—— (1993b) *Retailing in the European Single Market, 1993*, Brussels: Commission of the European Communities, 1993.

—— (1993c) *Labour Force Survey, 1983–1991*, Luxembourg, European Communities.

—— (1994) *Labour Force Survey, Results, 1992*, Luxembourg, European Communities.

—— (1995a) *Retailing in the European Economic Area*, Brussels: Commission of the European Communities.

—— (1995b) *Tourism in Europe*, Brussels: Commission of the European Communities.

—— (1996a) *Statistics in Focus, Regions, 1996 (2)*, Luxembourg: Eurostat.

—— (1996b) *Statistics in Focus (4)*, Luxembourg: Eurostat.

—— (1996c) *Labour Force Survey, Results, 1994*, Luxembourg: Eurostat.

Cecchini, P. (1988) *The European Challenge 1992: The Benefits of a Single Market*, Aldershot: Wildwood House.

CEDEFOP, European Centre for the Documentation of Occupational Training (1994) *Occupations in the Tourist Sector, A Comparative Analysis in Nine Community States*, Berlin: CEDEFOP

Centre for Economic Policy Research (1995) *Unemployment: Choices for Europe, Monitoring European Integration 5*, London: Centre for Economic Policy Research.

Champion, A.G. (ed.) (1989) *Counterurbanization: The Changing Pace of Population Deconcentration*, London: Edward Arnold.

—— (1994) 'Population change and migration in Britain since 1981: evidence for continuing deconcentration', *Environment and Planning A*

Champion, A.G. and Townsend, A.R. (1990) *Contemporary Britain, A Geographical Perspective*, London: Edward Arnold.

Champion A.G., Moennesland, J. And Vandermotten, C. (1996) 'The new regional map of Europe', *Progress in Planning* 46, 1: 89.

Chomsky, N. (1994) *World Orders, Old and New*, London: Pluto Press.

Christopherson, S. (1989) 'Flexibility in the US service economy and the emerging spatial division of labour', *Transactions, Institute of British Geographers* New Series 14, 2: 131–44.

—— (1996) 'The production of consumption: retail restructuring and labour demand in the USA', in N. Wrigley and M.S. Lowe (eds) *Retailing, Consumption and Capital*, London: Longman.

Church, A. and Stevens, M. (1994) 'Unionization and the urban–rural shift in employment', *Transactions, Institute of British Geographers* New Series 19, 1: 111–18.

Clark, C.A. (1940) *The Condition of Economic Progress*, London: Macmillan.

Clarke, I.M. (1985) *The Spatial Organisation of Multinational Corporations*, London: Croom Helm.

Clout, H. (1993) 'European experience of rural development', *Strategy Review: Topic Paper* 5, Salisbury: Rural Development Commission.

Coates, K. and Holland, S. (eds) (1995) *Full Employment for Europe*, London: Elf Books.

Cochrane, S.G. and Vining, D.R. (1986) 'Recent trends in migration between core and peripheral regions in developed and advanced developing countries', *Working Papers in Regional Science and Transportation* 108, Philadelphia: Regional Science Department, University of Pennsylvania.

Coe, N. (1996) 'Uneven development in the UK computer services industry since 1981', *Area* 28, 1: 64–77.

Cole, J. and Cole, F. (1993) *The Geography of the European Community*, London: Routledge.

Collier, J. (1994) 'Regional disparities, the Single Market and European Monetary Union', in J. Michie and J. Grieve-Smith (eds) *Unemployment in Europe*, London: Academic Press.

Commission on Social Justice, The (1994) *Social Justice: Strategies for National Renewal*, London: Vintage.

Conseil Economique et Social (1988), *Pour une Industrie Touristique plus Competitive*, Paris: Conseil Economique et Social.

Cooke, P. and Morgan, K.(1994), in A. Amin and N. Thrift (eds) *Globalisation, Institutions, and Regional Development in Europe*, Oxford: Oxford University Press.

Cooke, P., Moulaert, F., Swyngedouw, E., Weinstein, O. and Wells, P. (1992) *Towards Global Localisation: The Computing and Communications Industries in Britain and France*, London: University College London Press.

Cribier, F. (1982) 'Aspects of retirement migration from Paris: an essay in social and cultural geography', in A. M. Warnes (ed.) *Geographical Perspectives on the Elderly*, Chichester: John Wiley.

Cuadrado-Roura, J. and Gomez, C. (1992) 'Services and metropolitan centres: the expansion and location of business services', *Service Industries Journal* 12, 1: 97–115.

Curran, J. and Blackburn, R. (1994) *Small Firms and Local Economic Networks*, London: Paul Chapman.

Curran, J. and Stanworth, J. (1986) 'Trends in small firm industrial relations and their implications for the role of the small firm in restructuring', in A. Amin and J.B. Goddard (eds) *Technological Change, Industrial Restructuring and Regional Development*, Hemel Hempstead: Allen & Unwin.

Daly, M., Campbell, M., Robson, G. and Gallagher, C. (1991)' Job creation 1987–89: The contribution of small and large firms', *Employment Gazette* 99, 11: 589–96.

Daniels, P.W. (1985) *Service Industries: A Geographical Appraisal*, London: Methuen.

—— (1991) 'Producer services and the development of the space economy', in P. W. Daniels and F. Moulaert (1991) *The Changing Geography of Advanced Producer Services*, London: Belhaven.

—— (1993) *Service Industries in the World Economy*, Oxford: Blackwell.

Daniels, P.W. and Moulaert, F. (1991) *The Changing Geography of Advanced Producer Services*, London: Belhaven.

Daniels, P.W., Van Dinteren, J.H.J. and Monnoyer, M.C. (1991) *Business Services and the Urban Hierarchy in Western Europe*, Milan: European Research Network on Services and Space.

Davis, H.C. and Hutton, T.A. (1991) 'An empirical analysis of producer service exports from the Vancouver Metropolitan Region', *Canadian Journal of Regional Science* 14, 1: 89–97.

Davies, R.L. (ed.) (1995) *Retail Planning Processes in Western Europe*, London: Routledge.

Dawes, L. (1993) *Long-term Unemployment and Labour Market Flexibility*, Leicester: University of Leicester, Centre for Labour Market Studies.

Dawson, J.A. and Burt, S. (1989) *The Evolution of European Retailing*, Stirling: University of Stirling Institute for Retail Studies.

Deegan, J. and Dineen, D. (1992) 'Employment effects of Irish tourism projects', in P. Johnson and B. Thomas (eds) *Perspectives on Tourism Policy*, London: Mansell.

Dernoi, L.A. (1983) 'Farm tourism in Europe', *Tourism Management* 4, 1: 55–66.

Department of Employment (1994) *New Earnings Survey* London: HMSO

Department of the Environment, Ministry of Agriculture, Fisheries and Food (1995) *Rural England*, London: HMSO.

Dex, S. and Shaw, L.B. (1986) *British and American Women at Work*, London: Macmillan

Diamond, D.R. (1991) 'The City, the "Big Bang" and office development', in K. Hoggart and D.R. Green, *London: A New Metropolitan Geography*, London: Edward Arnold.

Dicken, P. (1986, first edition) *Global Shift: The Internationalization of Economic Activity*, London: Methuen.

—— (1990) 'European industry and global competition', in D. Pinder, (ed.) *Western Europe: Challenge and Change*, London: Belhaven.

—— (1992, second edition) *Global Shift: The Internationalization of Economic Activity*, London: Paul Chapman.

Dicken, P. and Oberg, S. (1996) 'The global context: Europe in a world of dynamic economic and population change', *European Urban and Regional Studies*, 3, 2: 101–20.

Dickinson, D. (1994) 'Crime and unemployment', unpublished paper by Cambridge Economist, cited in Report of the Commission on Social Justice, *Social Justice: Strategies for National Renewal*, London: Vintage.

Distributive Trades Economic Development Committee (1988) *Part-time Working in the Distributive Trades*, Vol. 1 *Training Practices and Career Opportunities*, London: National Economic Development Office.

Dreze, J.H. and Bean, C.R. (eds) (1990) *Europe's Unemployment Problem*, Cambridge, Mass: MIT Press.

Ducatel, K. and Blomley, N. (1990) 'Rethinking retail capital', *International Journal of Urban and Regional Research* 14, 2: 207–27.

Duke, R.C. (1993) 'European new entry into UK grocery retailing', *The International Journal of Retail and Distribution Management* 21, 1: 35–9.

Duncan, S. (1991) 'The geography of gender divisions of labour in Britain', *Transactions, Institute of British Geographers* New Series 16, 4: 420–40.

—— (1994) 'The diverse worlds of European patriarchy', *Environment and Planning A* 26, 8: 1174–6.

Dunford, M. (1995) 'Cohesion, growth, and inequality in the European Union', in A. Amin and J. Tomaney (eds) *Behind the Myth of European Union*, London: Routledge.

Dunning, J. (1988) *Multinationals and the European Community*, Oxford: Blackwell.

Echeverri-Carroll, E.L. (1994) 'Flexible linkages and offshore assembly facilities in developing countries', *International Regional Science Review* 17, 1: 49–74.

Economist Intelligence Unit (1993) *Europe's Top Retailers*, London: Economist Intelligence Unit.

Elias, P. and Hogarth, T. (1994) 'Families, jobs and unemployment: the changing pattern of economic dependency in Britain', in R. Lindley, (ed.) *Labour Market Structures and Prospects for Women*, Manchester: Equal Opportunities Commission.

Ellison, R. (1994) 'British labour force projections: 1994 to 2006', *Employment Gazette* 102, 4: 111–21

Elson, D. (1991) 'Male bias in macro-economics: the case of structural adjustment', in D. Elson (ed.) *Male Bias in the Development Process*, Manchester: Manchester University Press.

England, K.V.L. (1994) 'Blue-Collar Blues: Canadian Women's Manufacturing Employment and NAFTA', in Commission on Gender and Geography, International Geographical Union 1994: *Beyond Borders: The Gender Implications of Multistate Economic and Social Policies*, Conference proceedings, Heidelberg, 16–20 August.

Entorf, H., Franz, W., Koenig, H. and Smolny, W. (1990) 'The development of German employment and unemployment' in J.H. Dreze and C.R. Bean (eds) *Europe's Unemployment Problem*, Cambridge, Mass: MIT Press.

Esping-Andersen, G. (1990) *The Three Worlds of Welfare Capitalism*, Princeton, NJ: Princeton University Press.

Ethical Investment Research Service (EIRIS) (1990) 'Firms who are up in arms', *The Guardian* 29 January 14.

Etzioni, A. (1991), 'Socio-economics: a budding challenge' in A. Etzioni and P.R. Lawrence (eds) *Socio-Economics: Towards a new Synthesis*, Armonk, NY: Sharpe.

Fernie, J. (1995) 'The coming of the fourth wave', *International Journal of Retail & Distribution Management* 23, 1: 4–11.

Fielding, A.J. (1990) 'Counterurbanisation: threat or blessing?', in D. Pinder (ed.) *Western Europe: Challenge and Change*, London: Belhaven.

—— (1994) 'Industrial change and regional development in Western Europe', *Urban Studies* 31, 4/5: 679–704.

Finlay-Jones, R. and Eckhardt, B. (1981) 'Psychiatric disorder among the young unemployed', *Australian and New Zealand Journal of Psychiatry* 15, 3: 265–70.

Fisher, A.G.B. (1935) *The Clash of Progress and Security*, London: Macmillan.

Flanders, S. (1996) 'Life, jobs and the safety zone', *Financial Times* 29 April: 19.

Flanders, S. and Wolf, M. (1995) 'Haunted by the trade spectre', *Financial Times* 24 July: 13.

Foord, J., Bowlby, S. and Tillsley, C. (1992) 'Changing relations in the retail-supply chain, geographical and employment implications', *International Journal of Retail & Distribution Management* 20, 5: 23–30.

Forsberg, G. (1994) 'Occupational sex segregation in a "woman-friendly" society – the case of Sweden' *Environment and Planning A* 26, 8: 1235–56.

Fothergill, S. and Guy, N. (1990) *Retreat from the Regions: Corporate Change and the Closure of Factories*, London: Jessica Kingsley.

Fourastie, J. (1949) *Le Grand Espoir du XXE Siècle*, Paris: Presses Univ. de France

Franzmeyer, F., Hrubesch, P., Seidel, B., Weise, C. and Schweiger, I. (1991) *The Regional Impact of Community Policies*, Luxembourg: CEC.

Freeman, R.B. (1995) 'The limits of wage flexibility to curing unemployment', *Oxford Review of Economic Policy* 11, 1: 63–72.

Frey, W.H. (1993) 'The new urban revival in the United States', *Urban Studies* 30, 4/5: 741–74.

Gagey, F., Lambert, J.P. and Ottenwalelter, B. (1990) 'The development of German employment and unemployment: estimation and simulation of a small macro model' in J.H. Dreze and C.R. Bean (eds) *Europe's Unemployment Problem*, Cambridge, Mass: MIT Press.

Garcia-Ramon, M.D. and Baylina, M. (1994) 'Industrial homeworking and flexibility of the labour market in rural Spain', in Commission on Gender and Geography, International Geographical Union 1994: *Beyond Borders: The Gender Implications of Multistate Economic and Social Policies*, Conference proceedings, Heidelberg, 16–20 August.

Gentle, C. and Howells, J. (1994) 'The computer services industry: restructuring for a single market', *Tijdschrift voor Economische en Sociale Geographie* 85: 311–21.

Gershuny, J.I. (1979) 'The informal economy: its role in industrial society', *Futures* February: 3–15.

—— (1985) 'Economic development and change in the mode of provision of services', in N. Redclift and E. Mingione (eds) *Beyond Employment*, Oxford: Blackwell.

Gershuny, J. and Marsh, C. (1993) 'Unemployment in work histories', in D. Gallie, C. Marsh and C. Vogler (eds) *Social Change and the Experience of Unemployment*, London: Oxford University Press.

Gibb, R., and Michalak, W. (eds) (1994) *Continental Trading Blocs: The Growth of Regionalism in the World Economy*, Chichester: John Wiley.

Giddens, A. (1996) 'Affluence, poverty and the idea of a post-scarcity society', in C.H. de Alcantara (ed.) *Social Futures, Global Visions*, Oxford: Blackwell.

Giersch, H. (1985) 'Perspectives on the world economy', *Weltwirtschaftspolitik Archiv*, 121, 5: 409–26.

Giersch, H., Paque, K. and Schmieding, H. (1992) *The Fading Miracle: Four Decades of the Market Economy in Germany*, Cambridge: Cambridge University Press.

Gilg, A.W. (1991) 'Switzerland: structural change within stability', in A.M. Williams and G. Shaw (eds) *Tourism and Economic Development, Western European Experiences*, London: Belhaven.

Glasmeier, A. and Howland, M. (1994) 'Service-led rural development: definitions, theories, and empirical evidence', *International Regional Science Review* 16, 1/2: 187–96.

Glyn, A. (1995) 'The assessment: unemployment and inequality', *Oxford Review of Economic Policy* 11, 1: 1–25.

Goe, W. (1990) 'Producer services, trade and the social division of labour', *Regional Studies* 24, 4: 327–42.

Gold, M. (1993) *The Social Dimension*, London: Macmillan.

Goodhart, D. (1994) 'How to get the drop-outs back' in', *Financial Times* 4 December: 17.

Goss, J. (1992) 'Modernity and post-modernity in the retail landscape', in K. Anderson and F. Gale (eds) *Inventing Places: Studies in Cultural Geography*, Melbourne: Longman Cheshire.

Gottmann, J. (1961) *Megalopolis: the Urbanised Northeastern Seaboard the United States*, Cambridge, Mass: MIT Press.

Granovetter, M. (1985) 'Economic action and social structure: the problem of embeddedness' *American Journal of Sociology* 91, 3: 481–510.

Green, A. E., Owen, D.W. and Winnett, C.M. (1994) 'The changing geography of recession: analyses of local unemployment time series', *Transactions of the Institute of British Geographers* New Series 19, 2: 142–62

Gregg, P. (1993) 'Jobs and justice', in E. Balls and P. Gregg *Work and Welfare: Tackling the Jobs Deficit*, London: IPPR.

Gregg, P. and Wadsworth, J. (1995) 'A short history of labour turnover, job tenure, and job security', *Oxford Review of Economic Papers* 11, 73–89.

Gregory, D., Martin, R. and Smith, G. (eds) (1994) *Human Geography: Society, Space and Social Science*, London: Macmillan.

Gregson, N. and Lowe, M.(1994) *Servicing the Middle Classes*, London: Routledge.

Greenwood, J., Levy, R. and Stewart, R. (1995) 'The European Union Structural Fund Allocations', *European Urban and Regional Studies*, 2, 4: 317–38

Griffin, G., Wood, S. and Knight, J. (1992) 'The Bristol labour market', *Department of Employment Research Paper*, 82. Sheffield: Department of Employment.

Guerrier, Y. and Lockwood, A.J. (1989) 'Managing flexible working in hotels', *The Service Industries Journal* 9, 3: 406–19.

Hadjimichalis, C. and Sadler, D. (eds) (1995) *Europe at the Margins*, Chichester: John Wiley.

Hakim, C. (1988) 'Homeworking in Britain', in R.E. Pahl (ed.) *On Work: Historical, Comparative and Theoretical Approaches*, Oxford: Blackwell.

—— (1993) 'The myth of rising female employment', *Work, Employment and Society* 7, 1: 97–120.

Hale, D. (1995) 'Go slow ahead', *Financial Times* 3 January: 10.

Halimi, S., Michie, C. and Sefton, J. (1994) 'The Mitterrand Experience' in J. Michie and J. Grieve-Smith (eds) *Unemployment in Europe*, London: Academic Press.

Hall, P., Breheny, M., McQuaid, R. and Hart, D. (1989) *Western Sunrise*, London: Allen & Unwin.

Hamnett, C. (1994) 'Social polarisation in global cities: theory and evidence', *Urban Studies* 31, 3: 401–24.

Hannigan, K. (1994) 'A regional analysis of tourism growth in Ireland', *Regional Studies* 28, 2: 208–14.

Hanson, S. and Pratt, G. (1995) *Gender, Work and Space*, London: Routledge.

Hardy, S., Hart, M., Albrechts, L. and Katos, A. (1995) *An Enlarged Europe: Regions in Competition?*, London: Jessica Kingsley.

Harrington, J.W., Macpherson, A.D. and Lombard, J.R. (1991) 'Interregional trade in producer services: review and synthesis', *Growth and Change* 22: 75–94.

Harrop, A. and Moss, P. (1994) 'Working parents: trends in the 1980s', *Employment Gazette* 102, 10: 343–57.

Harrop, J. (1996) *Structural Funding and Employment in the European Union*, London: Elgar.

Harvie, C. (1994) *The Rise of Regional Europe*, London: Routledge.

Hasluck, C. and Green, A. (1992) 'Implications of the changing incidence of the long-term unemployed for the planning of programme provision and training', in E. McLaughlin (ed.) *Understanding Unemployment: New Perspectives on Active Labour Market Policies*, London: Routledge.

Helbrecht, I. and Pohl, J., (1993) 'Muenchen zwischen Expansion und Kollaps', *Geographische Rundschau* 45, 4: 238–43.

Hessels, M. (1994) 'Business services in the Randstad, Holland: decentralization and policy implications', *Tidjschrift voor Economische en Sociale Geographie* 85: 371–8.

Hicks, L. (1990) 'Excluded women: how can this happen in the hotel world?' *The Service Industries Journal* 10, 2: 348–63.

Hirst, P. and Zeitlin, J. (1992) 'Flexible specialisation versus post-fordism: theory, evidence and policy implications', in M. Storper and A.J. Scott (eds) *Pathways to Industrialisation and Regional Development*, London: Routledge.

Hobsbawm, E. (1994) *Age of Extremes: the Short Twentieth Century*, Micheal Joseph: London.

Hodges, L. (1995) 'Into the minds of management', *The Times Higher* 17 February: 17–18.

Howard, E. (ed.)(1990) *Leisure and Retailing*, Harlow: Longman.

—— (1992) 'Evaluating the success of out-of-town regional shopping centres', *The International Review of Retail, Distribution and Consumer Research* 2, 1: 59–80.

Howells, J. and Wood, M. (1993) *The Globalisation of Production and Technology*, London: Belhaven.

Howland, M. (1988) 'Plant closures and local economic conditions', *Regional Studies* 22, 3: 193–208.

Hudson, R. and Townsend, A. (1992) 'Tourism employment and policy choices in local government', in P. Johnson and B. Thomas (eds) *Perspectives on Tourism Policy*, London: Mansell.

Hudson, R., Schech, S. and Hansen, L. (1992) *Jobs for the Girls? The New Private Sector*

Economy of Derwentside District, Occasional Publication No. 28. Durham: Department of Geography, University of Durham.

Huigen, P. and Volkers, C.R. (1991) 'Counterurbanization: a process of regional rural development?', in P. Brunet (ed.) *Le Developpement Regional Rural en Europe*, Caen: Press Provinciale

Hutton, W. (1995) *The State We're In*, London: Jonathan Cape.

Huws, U. (1993) *Teleworking in Britain*, Research Series No. 18, London: Department of Employment.

Illeris, S. (1989) *Services and Regions in Europe*, Aldershot: Avebury.

—— (1991a) 'The many roads towards a service society', *Norsk Geografisk Tidskrift* 45, 1: 1–10.

—— (1991b) 'Location of services in a service society', in P.W. Daniels and F. Moulaert (1991) *The Changing Geography of Advanced Producer Services*, London: Belhaven.

—— (1994) 'Proximity between service producers and service users', *Tidjschrift voor Economische en Sociale Geographie* 85: 294–302.

Institute for Employment Research (1994) *Review of the Economy and Employment, Regional Assessment*, Warwick: Institute for Employment Research.

International Labour Organization (1995) *World Employment 1995*, Geneva: International Labour Organisation.

Investment Property Databank (1991) *IPD Annual Review*, London: IPD.

Jaeger, C. and Duerrenberger, G. (1991) 'Services and counterurbanisation: the case of central Europe', in P.W. Daniels (ed.) *Services and Metropolitan Development*, London: Routledge.

Jansen-Verbeke, M. (1990) 'Leisure + shopping = tourism product mix', in G. Ashworth and B. Goodall (eds) *Marketing Tourism Places*, London: Routledge.

Johnson, P. and Thomas, B. (1992) 'Tourism research and policy: an overview', in P. Johnson and B. Thomas (eds) *Perspectives on Tourism Policy*, London: Mansell.

Johnson, P. and Thomas, R.B. (1990) 'Employment in tourism: a review', *Industrial Relations Journal* 21, 10: 36–48.

Jones, A. (1994) *The New Germany: A Human Geography*, Chichester: John Wiley.

Joseph Rowntree Foundation (1995) *Inquiry into Income and Wealth*, York: Joseph Rowntree Foundation

Kassab, C. and Porterfield, S.L. (1991) 'The Spatial Allocation of Labour in the Producer Services Industries', Paper to the North American Regional Science Association, New Orleans, 7–10 November.

Kayser, B. (1990) *La Renaissance Rurale: Sociologie des Campagnes du Monde Occidental*, Paris: Arman Colin.

Keeble, D. and Tyler, P. (1995) 'Enterprising behaviour and the urban–rural shift', *Urban Studies* 32, 6: 975–97.

Keeble, D. and Wever, E. (eds) (1986) *New Firms and Regional Development in Europe*, London: Croom Helm.

Keeble, D., Bryson, J. and Wood, P. (1991) 'Small firms, business service growth and regional development in the United Kingdom: some empirical findings', *Regional Studies* 25, 5: 439–58.

Keeble, D., Tyler, P., Broom, G. and Lewis, J. (1992) *Business Success in the Countryside*, London: HMSO.

Kelly, T. (1987) *The British Computer Industry: Crisis and Development*, London: Croom Helm.

Kelly, T. and Keeble, D. (1990) 'IBM: the corporate chameleon', in M. de Smidt and E. Wever *The Corporate Firm in a Changing World Economy*, London: Routledge.

Kesteloot, C. (1995) 'Immigrant ghettoes in Brussels', in C. Hadjimichalis and D. Sadler *Europe at the Margins*, Chichester: John Wiley.

Kiernan, K. (1992) 'The role of men and women in tomorrow's Europe', *Employment Gazette* 100, 10: 491–8.

King, R. (1991) 'Italy: multi-faceted tourism', in A.M. Williams and G. Shaw (eds) *Tourism and Economic Development, Western European Experiences*, London: Belhaven.

—— (1992) 'Italy: from sick man to rich man of Europe', *Geography* 2: 153–69.

Kinnaird, V. and Hall, D. (1994) (eds) *Tourism: A Gender Analysis*, Chichester: John Wiley.

Kreisky Commission (1989) *A Programme for Full Employment in the 1990s*, Oxford: Pergamon.

Krugman, P. (1990) *The Age of Diminished Expectations: US Economic Policy in the 1990s*, Washington: Washington University Press.

Kruse, W., Bertrand, O., Homs, O., Mendez-Vigo, M. and Van den Tillaart, H. (1993) *Training in the Retail Sector: A Survey for the Force Programme*, Berlin: CEDEFOP.

Labour Force Survey, Quarterly Bulletin. London: Department of Employment Group to summer, 1995 then Central Statistical Office.

Laulajainen, R., Abe, K. and Laulajainen, T. (1993) 'The geographical dimension of global retailing', *The International Review of Retail, Distribution and Consumer Research* 3, 4: 367–90.

Law, C.M. (1994) *Urban Tourism: Attracting Visitors to Large Cities*, London: Mansell.

Law, C.M. and Warnes, A.M. (1982) 'The destination decision in retirement migration', in A.M. Warnes (ed.) *Geographical Perspectives on the Elderly*, Chichester: John Wiley.

Layard, R., Nickell, S. and Jackman, R. (1994) *The Unemployment Crisis*, Oxford: Oxford University Press.

Lentnek, B., MacPherson, A. and Philips, D. (1994) 'Optimum producer-service location', *Environment and Planning A* 26: 467–79.

Leontidou, L. (1994) 'Gender dimensions of tourism in Greece: employment, subcultures and restructuring', in V. Kinnaird and D. Hall (eds) *Tourism: A Gender Analysis*, Chichester: John Wiley.

Leyshon, A. and Thrift, N. (1989) 'South goes north? The rise of the British provincial financial centre', in J. Lewis and A. Townsend (eds) *The North–South Divide*, London: Paul Chapman.

Lindley, R. (ed.) (1992) *Women's Employment: Britain in the Single European Market*, Manchester: Equal Opportunities Commission.

—— (1994) 'Economic and social dimensions', in R. Lindley (ed.) *Labour Market Structures and Prospects for Women*, Manchester: Equal Opportunities Commission.

Lindley, R. and Wilson, R. (1992) 'SEM scenarios for the employment of women and men in Great Britain', in R. Lindley (ed.) *Women's Employment: Britain in the Single European Market*, Manchester: Equal Opportunities Commission.

—— (eds) (1993) *Review of the Economy and Employment: Occupational Assessment*, Coventry: Institute for Employment Research, University of Warwick.

—— (1992) *Towards a New Economic Order: Post-Fordism, Ecology and Democracy*, New York: Polity Press.

—— (1993) Statements reported from the oral version of conference papers later appearing as, 'The local and the global: regional individuality and inter-regionalism', *Transactions, Institute of British Geographers* New Series 18, 1: 8–18.

Lipietz, A. (1992) *Towards a New Economic Order; Post-Fordism, Ecology and Democracy*, New York: Polity Press.

—— (1993) Statements reported from the oral version of conference papers later appearing as, 'The local and the global: regional individuality and inter-regionalism', *Transactions, Institute of British Geographers* New Series 18, 8–18.

Little, J. (1991) 'Women in the rural labour market: a policy evaluation', in T. Champion and C. Watkins *People in the Countryside*, London: Paul Chapman.

Loesch, A. (English translation) (1954) *Economics of Location*, New Haven: Yale University Press.

Lowe, M.S. (1991) 'Trading places: retailing and local economic development at Merry Hill, West Midlands', *East Midlands Geographer* 14, 1: 31–48.

Lowe, M.S. and Short, J.R. (1990) 'Progressive human geography', *Progress in Human Geography* 14, 1: 1–11.

Lowe, M.S. and Wrigley, N. (eds) (1996) *Retailing, Consumption and Capital*, Harlow: Longman.

Macdonald, R. (1996) 'Urban Tourism in Liverpool: Evidence from Providers, the Case of Liverpool', unpublished Ph.D. thesis, University of Durham.

McDowell, L. (1991) 'Life without father and Ford: the new gender order of post-Fordism', *Transactions, Institute of British Geographers* New Series 16, 4: 400–19.

McDowell, L. and Court, G. (1994a) 'Performing work: bodily representations in merchant banks', *Environment and Planning D: Society and Space* 12, 12: 727–50.

—— (1994b) 'Gender divisions of labour in the post-Fordist economy: the mainten-ance of occupational sex segregation in the financial services sector', *Environment and Society A* 26, 9: 1397–1418.

McGregor, A. and Sproull, A. (1992) 'Employers and the flexible workforce', *Employ-ment Gazette* 100, 5: 225–34.

Marsh, A. and Mackay, S. (1993) 'Families, work and the use of childcare', *Employment Gazette* 101, 8: 361–70.

Marshall, J.N. (1985) 'Business services, the regions and regional policy', *Regional Studies* 194: 353–63.

—— (1994) 'Business reorganization and the development of corporate services in metropolitan areas', *The Geographical Journal* 160, 1: 41–9.

Marshall, J.N. and Jaeger, C. (1990) 'Service activities and uneven spatial development in Britain and its European partners: deterministic fallacies and new options', *Environment and Planning A* 22, 10: 1337–54.

Marshall, J.N. and Richardson, R. (1996) 'The impact of "telemediated" services on corporate structures: the example of "branchless" banking in Britain', *Environment and Planning A* 28, 10: 1843–58.

Marshall, J.N. and Wood, P.A. (1992) 'The role of services in urban and regional development', *Environment and Planning A* 24, 9: 1255–70.

—— (1995) *Services and Space: Key Aspects of Urban and Regional Development*, Harlow: Longman.

Marshall, J.N. *et al.* (1988) *Services and Uneven Development*, Oxford: Oxford University Press.

Marshall, M. (1987) *Long Waves of Regional Development*, London: Macmillan.

Martin, J. and Roberts, C. (1984) *Women and Employment: A Lifetime Perspective*, London: HMSO.

Martin, R. (1994) 'Economic theory and human geography', in D. Gregory, R. Martin and G. Smith (eds) *Human Geography: Society, Space and Social Science*, Basingstoke: Macmillan.

Martin, R. and Rowthorn, B. (eds) (1986) *The Geography of Deindustrialisation*, Basing-stoke: Macmillan.

Martinelli, F. (1991) 'Branch plants and services underdevelopment in peripheral regions: the case of southern Italy', in P.W. Daniels and F. Moulaert (1991) *The Changing Geography of Advanced Producer Services*, London: Belhaven.

Masser, I., Sviden, O. and Wegener, M. (1992) *The Geography of Europe's Futures*, Lon-don: Belhaven.

Massey, D. (1983) 'Industrial restructuring as class restructuring: production decentral-isation and local uniqueness', *Regional Studies* 17, 1: 73–90.

—— (1984) *Spatial Divisions of Labour*, Basingstoke: Macmillan.

Massey, D. and Meegan, R. (1988) 'What's happening to UK manufacturing', in J. Allen and D. Massey (eds) *Restructuring Britain: The Economy in Question*, London: Sage.

Matzner, E. and Streeck, W. (1991) 'Introduction: towards a socio-economics of employment in a post-Keynesian economy', in E. Matzner and W. Streeck *Beyond Keynesianism*, Aldershot: Elgar.

Mayere, A. and Vinot, F. (1991) 'Service firms' networks and characteristics of clientele', paper presented at the thirty-first European Congress of the Regional Science Association, Lisbon, Portugal, August.

Meadows, P. (1996) (ed.) *Work out – or Work in? Contributions to the debate on the future of work*, York: Rowntree Foundation.

Medlik, S. (1988) *Tourism and Productivity*, London: BTA/ETB Research Services.

Meulders, D., Plasman, R. and Stricht, V. (1993) *Position of Women on the Labour Market in the European Community*, Aldershot: Dartmouth.

Meulders, D., Plasman, O. and Plasman, R. (1994) *Untypical Employment in the EC*, Aldershot: Dartmouth.

Michie, J. and Grieve-Smith, J. (eds) (1994) *Unemployment in Europe*, London: Academic Press.

Mincer, J. (1962) 'Labour force participation of married women: a study of labor supply', in National Bureau of Economic Research, *Aspects of Labor Economics: A Conference of the Universities*, Princeton, NJ: Princeton University Press.

Minford, P. (1995) 'The elixir of growth: trade, non-traded goods and development', *Centre for Economic Policy Research Discussion Paper*, London: Centre for Economic Policy Research.

Mingione, E. (1988) 'Work and informal activities in urban Southern Italy', in R.E. Pahl *On Work: Historical, Comparative and Theoretical Approaches*, Oxford: Blackwell.

—— (1993) 'The new urban poverty and the underclass: Introduction', *International Journal of Urban and Regional Research* 17, 3: 324–6.

Mingione, E. and Morlicchio, E. (1993) 'New forms of urban poverty in Italy: risk path models in the North and South', *International Journal of Urban and Regional Research* 17, 3: 413–28.

Monnoyer, M.-C. and Philippe, J. (1991) 'Localisation factors and development strategies of producer services', in P. W. Daniels and F. Moulaert (1991) *The Changing Geography of Advanced Producer Services*, London: Belhaven.

Morgan, G. (1992) 'The globalisation of personal financial services: the European Community after 1992', *Service Industries Journal* 12, 2: 193–209.

Morris, L.D. (1993) 'Is there a British underclass?,' *International Journal of Urban and Regional Research* 17, 3: 404–13.

Morris, W. (1995) 'Introduction', in *Economic Policies for Full Employment and the Welfare State*, London, PO Box 188, SW1A 0SGA: Campaign to Defend the Welfare State/Full Employment Forum pamphlet.

Moulaert, F. and Toedtling, F. (1995), 'The geography of advanced producer services in Europe – conclusions and prospects', *Progress in Planning* 43, 2/3: 1261–74.

Muenchau, W. (1996) 'A country losing competitiveness', *Financial Times* 8 February: 15

Munday, M., Morris, J. and Wilkinson, B. (1995) 'Factories or warehouses? A Welsh perspective on Japanese transplant manufacturing', *Regional Studies* 29, 1: 1–18.

Musterd, S. (1994) 'A rising European underclass? Social polarization and spatial segregation in European cities', *Built Environment* 20, 3: 85–191.

National Children's Home (1993) *British Children in Need 1993*, London: National Children's Home.

Naylor, K. (1994) 'Part-time working in Great Britain – an historical analysis', *Employment Gazette* 102, 12: 472–84.

NEDO (National Economic Development Office) (1986) *Changing Working Patterns: How Companies Achieve Flexibility to Meet New Needs*, London: National Economic Development Office.

Newby, H. (1980) *Green and Pleasant Land? Social Change in Rural England*, Harmondsworth: Penguin.

Nickell, S. and Bell, B. (1995) 'The collapse in demand for the unskilled and unemployment across the OECD', *Oxford Review of Economic Policy* 11, 1: 40–62

Norman, P. (1995) 'Mood swings cloud German recovery', *Financial Times*: 1 September: 11

Noyelle, T.J. and Stanback, T.M. (1984) *The Economic Transformation of American Cities*, Totowa, NJ: Rowman & Allanheld.

OECD (1961) *Labour Force Statistics, 1956–1966*, Paris: OECD.

—— (1986) *Tourism Policy and International Tourism in OECD Countries*, Paris: OECD.

—— (1994a) *Labour Force Statistics, 1972–1992*, Paris: OECD.

—— (1994b) *The OECD Jobs Study: Evidence and Explanations*, Paris: OECD

—— (1995) *Labour Force Statistics, 1973–1993*, Paris: OECD.

OECD Economic Surveys (1992) *United States*, Paris: OECD.

O'Farrell, P.N., Hitchens, D.M. and Moffatt, L.A.R. (1992) 'The competitiveness of business service firms in Scotland and the South-East of England: a matched pairs analysis', *Regional Studies* 26, 6: 519–33.

O'Farrell, P.N., Moffat, L.A.R. and Hitchens, D.M.W.N. (1993) 'Manufacturing demand for business services in a core and peripheral region: does flexible production imply vertical disintegration of business services?', *Regional Studies* 27, 5: 385–400.

O'Farrell, P.N., Wood, P.A. and Zheng, J. (1994) 'Internationalisation of business services: some preliminary evidence'. Paper given to the IVth Annual Reser Conference, Barcelona, September 22/23.

Ohmae, K. (1985) *Triad Power: The Coming Shape of Global Competition*, New York: Free Press.

O'hUallachain, B. and Reid, N. (1991) 'The location and growth of business and professional services in American metropolitan areas, 1976–1986', *Annals, Association of American Geographers* 81, 2: 254–70.

PA CEC (1988) *A Study of Rural Tourism*, PA Cambridge Economic Consultants Ltd, London: HMSO.

Pahl, R.E. (1984) *Divisions of Labour*, Oxford: Blackwell.

Parker, G. (1988) 'Who cares? A review of empirical evidence from Britain', in R.E. Pahl *On Work: Historical, Comparative and Theoretical Approaches*, Oxford: Blackwell.

Paukert, L. (1984) *Women's Employment and Unemployment in the OECD Area*, Paris: OECD.

Pearce, D.G. (1988) 'Tourism and regional development in the European Community', *Tourism Management* 8, 1: 13–22.

Peck, F. and Townsend, A.R. (1984) 'Contrasting experience of recession and spatial restructuring: British Shipbuilders, Plessey and Metal Box', *Regional Studies* 18, 4: 319–38.

—— (1987) 'The impact of technological change upon the spatial pattern of UK employment within major corporations', *Regional Studies* 21, 3: 225–40.

Peck, J. (1994) 'Regulating labour: the social regulation and reproduction of local labour markets', in A. Amin and N. Thrift (eds) *Globalization, Institutions and Regional Development in Europe*, Oxford: Oxford University Press.

Penn, R. (1995) 'Flexibility, skill and technical change in UK retailing', *Service Industries Journal* 15, 3: 229–42.

Penn, R. and Wirth, B. (1993) 'Employment patterns in contemporary retailing: gender and work in five supermarkets', *Service Industries Journal* 13, 4: 252–66.

Perrons, D. (1994) 'Measuring equal opportunities in European employment', *Environment and Planning A* 268: 1195–220.

Persky, J. and Wiewel, W. (1994) 'The growing localness of the global city', *Economic Geography* 70, 1: 129–43.

Pfau-Effinger, B. (1994) 'The gender contract and part-time paid work by women – Finland and Germany compared', *Environment and Planning A* 26, 9: 1355–76.

Phillips, K. (1990) *The Politics of Rich and Poor: Wealth and the Electorate in the Reagan Aftermath*, New York: Random House.

Piachaud, D. (1991) 'Unemployment and Poverty', *Campaign for Work, Research Report*, 3 (3), London: Tottenham Town Hall.

Pinch, S. and Storey, A. (1992) 'Flexibility, gender and part-time work: evidence from a survey of the economically active', *Transactions, Institute of British Geographers* New Series 17, 2: 198–214.

Piore, M. and Sabel, C. (1984) *The Second Industrial Divide: Possibilities for Prosperity*, New York: Basic Books.

Plane, D.A. (1989) 'Population migration and economic restructuring in the United States', *International Regional Science Review* 12, 3: 263–81.

Pratt, G. and Hanson, S. (1991) 'On the links between home and work: family-household strategies in a buoyant labour market', *International Journal of Urban and Regional Research* 15, 1: 55–74.

Quack, S. and Maier, F. (1994) 'From state socialism to market economy - women's employment in East Germany', *Environment and Planning A* 26, 8: 1257–96.

Ramsay, H. (1993) 'Labour and the Single Market', *Global Labour* 1, 1: 7–9.

Redclift, N. and Mingione, E. (1985) *Beyond Employment: Household, Gender and Subsistence*, Oxford: Blackwell.

Reich, R.B. (1991) *The Work of Nations*, London: Simon & Schuster.

Reichenbach, H., Muro, G.D. and Lehner, S. (1994) 'European integration and growth', in J. Michie and J. Grieve-Smith (eds) *Unemployment in Europe*, London: Academic Press.

Report of the Commission on Social Justice (1994) *Social Justice: Strategies for National Renewal*, London: Vintage.

Resource Center Bulletin (1993) 'Industry on the run', *Resource Center Bulletin* Albuquerque: Inter-Hemispheric Education Resource Center, 33, fall.

Reynolds, J. (1983) 'Retail employment change: scarce evidence in an environment of change', *Service Industries Journal* 3, 3: 334–62.

Reynolds, P., Storey, D.J. and Westhead, P. (1994) 'Cross-national comparisons of the variation in new firm formation rates', *Regional Studies* 28, 4: 443–56.

Rifkin, J. (1996) *The End of Work*, New York: Putnam.

Roberts, K. (1995) *Youth and Employment in Modern Britain*, Oxford: Oxford University Press.

Robinson, T.M. and Clarke-Hill, C.M. (1990) 'Directional growth by European retailers', *International Journal of Retail and Distribution Management* 18, 5: 3–13.

Rodgers, G. (1989) 'Precarious work in Western Europe: The state of the debate', in G. Rogers and J. Rodgers (eds) *Precarious Jobs in Labour Market Regulation: The Growth of Atypical Employment in Western Europe*, Geneva: International Labour Organisation.

Rodgers, G. and Rogers J. (eds)(1989) *Precarious Jobs in Labour Market Regulation: The growth of atypical employment in Western Europe*, Geneva: International Labour Organisation.

Rodwin, L. (1989) 'Deindustrialization and regional economic transformation', in L. Rodwin and H. Sazanami (eds) *Deindustrialization and Regional Economic Transformation*, London: Unwin Hyman.

Rowbotham, S. and Mitter, S. (1994) *Dignity and Daily Bread*, London: Routledge.

Rowley, G. (1993) 'Prospects for the central business district', in R.D.F. Bromley and C.J. Thomas (eds) *Retail Change, Contemporary Issues*, London: UCL Press.

Rubery, J. and Fagan, C. (1994) 'Occupational segregation: plus ça change . . . ?' in R. Lindley (ed.) *Labour Market Structures and Prospects for Women*, Manchester: Equal Opportunities Commission.

Rudolf, H., Appelbaum, E. and Maier, F. (1994) 'Beyond socialism: the uncertain prospects for East German women in a unified Germany', in N. Aslanbeigui, S. Pressman and G. Summerfield (eds) *Women in the Age of Economic Transformation*, London: Routledge.

Rural Development Commission, North, D. and Smallbone, D. (1993) 'Small business in rural areas', *Strategy Review: Topic Paper* 2, London: Rural Development Commission.

Sackmann, R. and Haussermann, H. (1994) 'Do regions matter? Regional differences in female labour-market participation in Germany', *Environment and Planning A* 26, 9: 1397–1418.

Sadler, D. (1992) 'Beyond "1992": the evolution of European Community policies towards the automobile industry', *Environment and Planning C: Government and Policy* 10, 2: 229–48

Sassen, S. (1991) *The Global City*, Princeton, NJ: Princeton University Press.

Sayer, A. and Walker, R. (1992) *The New Social Economy*, Oxford: Blackwell.

Sayers, S. (1988) 'The need to work: a perspective from Philosophy', in R.E. Pahl (ed.) *On Work: Historical, Comparative and Theoretical Approaches*, Oxford: Blackwell.

Scott, A.M. (ed.) (1994) *Gender Segregation and Social Change*, Oxford: Oxford University Press.

Sexton, P.C. (1991) *The War on Labor and the Left: Understanding America's Unique Conservatism*, Boulder, CO: Western Press.

Shaw, G. and Williams, A.M. (1988, first edition) 'Western European tourism in perspective', in A.M. Williams and G. Shaw (eds) *Tourism and Economic Development, Western European Experiences*, London: Belhaven.

—— (1990) 'Tourism and development', in D. Pinder (ed.) 1990: *Western Europe: Challenge and Change*, London: Belhaven.

—— (1994) *Critical Issues in Tourism: A Geographical Perspective*, Oxford: Blackwell.

Sidwell, E. (1979) 'London to Bristol: the experience of a major office decentralization', in P.W. Daniels (ed.) *Spatial Patterns of Office Growth and Location*, Chichester: John Wiley.

Sinden, A. (1995) 'The decline, flexibility and geographical restructuring of employment in UK retail banks, 1989–1993', *Geographical Journal* 162, 1: 25–40.

Singh, A. (1977) 'UK industry and the world economy: a case of deindustrialisation?', *Cambridge Journal of Economics* 1, 2: 113–362.

Sjoholt, P. (1994) 'Economic crisis and the tertiary sector: some evidence from the Scandinavian world'. Paper to the IVth Annual RESER Conference, Barcelona, 22/23 September.

Small Business Research Centre, University of Cambridge (1992) *The State of British Enterprise*, Cambridge: Small Business Research Centre.

Smith, J. (1987) 'Transforming households: working class women and economic crisis', *Social Problems* 34, 5: 416–36.

Smith, J.P and Ward, M.P. (1980) 'Time series growth in the female labor force', *Journal of Labor Economics* 3, 1: 59–90.

Sparks, L. (1983) 'Employment characteristics of superstore retailing', *Service Industries Journal* 3, 1: 63–78.

—— (1992) 'Restructuring retail employment', *International Journal of Retail and Distribution Management* 20, 3: 12–19.

Stanworth, J. (1993) *Self-employment and Labour Market Restructuring – The Case of Freelance Teleworkers in Book Publishing*, London: Westminster University.

Storey, D. and Johnson, P. (1987) *Job Generation and Labour Market Change*, Basingstoke; Macmillan.

Strambach, S. (1994) 'Knowledge-intensive business services in the Rhine-Neckar area', *Tidjschrift voor Economische en Sociale Geographie* 85: 354–65.

Stratigaki, M. and Vaiou, D. (1994) 'Women's work and informal activities in Southern Europe', *Environment and Planning A* 26, 8: 1221–34.

Streeck, W. (1987) 'The uncertainties of management in the management of uncertainty: employers, labor relations and industrial adjustment', *Work, Employment and Society* 1, 3: 281–308.

Stubbs, C. and Wheelock, J. (1990) *A Woman's Work in the Changing Local Economy*, Aldershot: Avebury.

Symes, V. (1995) *Unemployment in Europe*, London: Routledge.

Symonds, M. (1994) *The Culture of Anxiety: The Middle Class Crisis*, London: The Social Market Foundation.

Taylor, P. (1991) 'Unemployment and health', *Campaign for Work, Research Report* 3, 6, London: Tottenham Town Hall.

Thomas, A., with Crow, B., Frenz, P., Hewitt, T., Kassam, S. and Treagust, S. (second edition, 1994) *Third World Atlas*, Buckingham: The Open University.

Thomas, C.J. and Bromley, R.D.F. (1993) 'The impact of out-of-centre retailing', in R.D.F. Bromley and C.J. Thomas (eds) *Retail Change, Contemporary Issues*, London: UCL Press.

Thomas, R. (ed.) (1996) *The Hospitality Industry, Tourism and Europe*, London: Cassell.

Thrift, N.J. (1989) 'The geography of international economic disorder', in R.J. Johnston and P.J. Taylor (eds, second edition) *A World in Crisis? Geographical Perspectives*, Oxford: Blackwell.

Thrift, N., Leyshon, A. and Daniels, P. (1987) 'Sexy greedy: the new international system, the City of London and the South-East of England', *Working Papers on Financial Services 8* Bristol: Department of Geography, University of Bristol.

Tomaney, J. (1994) 'Regional and industrial aspects of unemployment in Europe', in J. Michie and J. Grieve-Smith (eds) *Unemployment in Europe*, London: Academic Press.

Tordoir, P.P. (1994) 'Transactions of professional business services and spatial systems', *Tidjschrift voor Economische en Sociale Geographie* 85: 322–32.

Townsend, A.R. (1983) *The Impact of Recession, on Industry, Employment and the Regions*, London: Croom Helm.

—— (1986) 'Spatial aspects of the growth of part-time employment in Britain', *Regional Studies* 20, 4: 313–30.

—— (1991a) 'Services and local economic development', *Area* 23, 4: 309–17.

—— (1991b) 'New forms of employment in rural areas: a national perspective', in T. Champion and C. Watkins (eds) *People in the Countryside*, London: Paul Chapman.

—— (1992a) 'New directions in the growth of tourism employment?: propositions of the 1980s', *Environment and Planning A* 24, 6: 821–32.

—— (1992b) 'Gender at work in the OECD area: new trends in female and part-time employment', *International Geographical Union, Study Group on Gender and Geography* Working Paper 25.

—— (1993) *Uneven Regional Change in Britain*, Cambridge: Cambridge University Press.

—— (1994a) 'Regional structures and perspectives', in R. Lindley (ed.) *Labour Market Structures and Prospects for Women*, Manchester: Equal Opportunities Commission.

—— (1994b) 'The urban–rural cycle in the Thatcher growth years', *Transactions, Institute of British Geographers* New Series 18, 2: 207–21

Townsend, A.R. and Macdonald, R. (1993) 'Business services in the economic development of Edinburgh', *Department of Geography, University of Durham*, Working Paper 1, Durham: Department of Geography.

—— (1995) 'Sales of services from a European city', *Tidjschrift voor Economische en Sociale Geogaphie* 86, 5: 443–55.

Townsend, A.R. and Peck, F.W. (1985) 'The geography of mass redundancy in named corporations', in M. Pacione (ed.) *Progress in Industrial Geography*, London: Croom Helm.

Townsend, A., Sadler, D. and Hudson, R. (1995) 'Regional employment trends in retailing in a "Two Nations" Britain', in N. Wrigley and M. Lowe (eds) *Retailing, Consumption and Capital: Towards the New Economic Geography of Retailing*, London: Longman.

Treadgold, A.D. (1990) *The Costs of Retailing in Continental Europe*, Harlow: Longman.

Treadgold, A.D. and Davies, R.L. (1988) *The Internationalisation of Retailing*, Harlow: Longman.

Trubowitz, P. and Roberts, B.E. (1992) 'Regional Interests and the Reagan military build-up', *Regional Studies* 26, 6: 555–68.

Tucker, E. and Tett, G. (1996) 'Belgium sends distress signals', *Financial Times*: 5 May 3

Tuppen, J.N. (1985) 'Urban tourism in France – a preliminary assessment', *Urban Tourist Project*, Working Paper 3, Salford: Department of Geography, University of Salford.

—— (1991) 'France: the changing character of a key industry', in G. Shaw and A.M. Williams (eds) 'Western European tourism in perspective', in A.M. Williams and G.Shaw (eds) *Tourism and Economic Development, Western European Experiences*, London: Belhaven.

Tuppen, J.N. and Thompson, I.B. (1994) 'Industrial restructuring in contemporary France: spatial priorities and policies', *Progress in Planning* 42, 2: 103–72.

UNCTAD, United Nations Committee for Trade and Development (1994) *World Investment Report Where*, UNCTAD.

UNIDO, United Nations Industrial Development Organization (1995) *Industrial Development Global Report, 1995*, Vienna: UNIDO.

Urry, J. (1990) *The Tourist Gaze*, London: Sage.

Valenzuela, M. (1991) 'Spain: the phenomenon of mass tourism', in A.M. Williams, and G. Shaw (eds) *Tourism and Economic Development, Western European Experiences*, London: Belhaven.

Van Dinteren, J.H.J. and Meuwissen, J.A.M. (1994) 'Business services in the core area of the European Union', *Tidjschrift voor Economische en Sociale Geographie* 85: 366–70.

Van Ritbergen, T. (1994) 'The insurance sector in geographical perspective', *Tidjschrift voor Economische en Sociale Geographie* 85: 263–8.

Vickers, J. and Wright, V, (1989) *The Politics of Privatisation in Western Europe*, London: Frank Cass.

Vining, D.R. and Kontuly, T. (1978) 'Population dispersal from major metropolitan regions: an international comparison', *International Regional Science Review* 3, 1: 49–73.

Walby, S. (1986) *Patriarchy at Work*, Cambridge: Polity Press.

—— (1990) *Theorizing Patriarchy*, Oxford: Blackwell.

Walsh, J.A. (1991) 'The turnaround of the turnaround in the population of the Republic of Ireland', *Irish Geography* 24, 2: 117–25.

Walsh, T. (1992) 'European economic integration: employment effects in the distributive trades', in R.M. Lindley (ed.) *Women's Employment: Britain in the Single Market*, London: HMSO.

Waring, M. (1989) *If Women Counted: A New Feminist Economics*, London: Macmillan.

Watson, G. (1994) 'The flexible workforce and patterns of working hours in the UK', *Employment Gazette* 102, 7: 239–47.

Webb, S. (1994) 'Women and the UK income distribution: past, present and prospects', in R. Lindley (ed.) *Labour Market Structures and Prospects for Women*, Manchester: Equal Opportunities Commission.

Weber, A. (English translation with notes by Friedrich) (1929) *Alfred Weber's Theory of the Location of Industries*, Chicago: Chicago University Press.

Whitebloom, S. and Springett, A. (1994) 'Plunderers are at the gates', *The Guardian* 22 January: 38.

Williams, A.M. and Shaw, G. (1991, second edition) 'Western European tourism in perspective', in A.M. Williams, and G. Shaw (eds) *Tourism and Economic Development, Western European Experiences*, London: Belhaven.

Williams, C.C. and Windebank, J. (1994) 'Spatial variations in the informal sector: a review of evidence from the European Union', *Regional Studies* 28, 8: 819–26.

Wilson, R.A. (1994) 'Sectoral and occupational change: prospects for women's employment', in R. Lindley (ed.) *Labour Market Structures and Prospects for Women*, Manchester: Equal Opportunities Commission.

Wise, M. and Gibb, R. (1993) *Single Market to Social Europe*, Harlow: Longman.

Wood, A. (1994) *North–South Trade, Employment and Inequality: Changing Fortunes in a Skill-Driven World*, Oxford: Clarendon Press.

Wood, P. A. (1991a) 'Flexible accumulation and the rise of business services', *Transactions, Institute of British Geographers* New Series 16, 2: 160–72.

—— (1991b) 'Conceptualising the role of services in economic change', *Area* 23, 1: 66–72.

Wood, S. (1989) 'The transformation of work', in S. Wood (ed.) *The Transformation of Work*, London: Routledge.

World Bank (1995) *World Development Report: Workers in an Integrating World*, New York: Oxford University Press.

World Tourism Organization (1993) *Tourism Trends Worldwide and in Europe, 1980–92*, Madrid: World Tourism Organization.

Wrigley, N. (1993) 'Retail concentration and the internationalization of British grocery retailing', in R.D.F. Bromley and C.J. Thomas (eds) *Retail Change: Contemporary Issues*, London: UCL Press.

Yzewyn, D. and De Brabander, G. (1992) 'The economic impact of tourism and recreation in the Province of Antwerp, Belgium', in P. Johnson and B. Thomas (eds) *Perspectives on Tourism Policy*, London: Mansell.

INDEX